Advance Praise for

Solar Electricity Basics

Solar Electricity Basics, is an indispensable primer for any homeowner or small business owner considering a photovoltaic system. As the title promises, Dan covers all the basics but also comprehensively addresses the nuts and bolts of every essential sub-topic. He has provided a highly readable and immediately useful addition to the body of solar literature.

— Ann Edminster, author,
Energy Free: Homes for a Small Planet

Dan Chiras clearly communicates theory and practice of a resource critically important to integrate into our national energy mix. Homeowners, students, municipal officials and business people can readily gain a solar working knowledge base from this thorough, accessible book.

— Jean Ponzi, Green Resources Manager,
EarthWays Center of Missouri Botanical Garden

Dan Chiras has done it again. In *Solar Electricity Basics*, he tackles a timely, vast, and highly technical subject in his typically thorough, clear and concise approach. From a look at the sun itself, through every component of a solar electric system, through design, installation and even permitting issues, he leaves no aspect of solar electricity out, including helping one to decide whether one is actually ready to venture off the grid. Richly illustrated and explained in easy-to-understand language, Chiras has provided everything one needs to know about generating electricity from the sun. I highly recommend this book!

— James Plagmann, AIA + LEED AP - Green Architect

Both accessible to those new to solar electric systems and a well-referenced review for experts, *Solar Electricity Basics* is chock full of practical and useful information. We already recommend *Power from the Sun* to all our new members and use it as supplemental material for our workshops. We look forward to doing the same with *Solar Electricity Basics*.

— Jeffrey Owens, Executive Director, Show Me Solar (ShowMeSolar.org)

Dan Chiras strikes again! His books — which occupy a whole shelf in my bookcase — have now enabled thousands of energy activists to convert sun, wind, and water into self-reliance. Even if you can't figure out how your TV remotes work, with *Solar Electricity Basics* as a guide, you can design, purchase, and install the perfect photovoltaic system for your home or business.

— David Wann, coauthor, *Affluenza* and *Superbia!*, and author, *Simple Prosperity* and *The New Normal*

SOLAR
ELECTRICITY
Basics

DAN CHIRAS

with Robert Aram and Kurt Nelson, Technical Advisors,
and Anil Rao, Ph.D., illustrator

NEW SOCIETY PUBLISHERS

Printed in Canada by Friesens. First printing May 2010.

Paperback ISBN: 978-0-86571-618-6

Inquiries regarding requests to reprint all or part of *Solar Electricity Basics*
should be addressed to New Society Publishers at the address below.
To order directly from the publishers, please call toll-free (North America)
1-800-567-6772, or order online at www.newsociety.com

Any other inquiries can be directed by mail to:

New Society Publishers
P.O. Box 189, Gabriola Island, BC V0R 1X0, Canada
(250) 247-9737

New Society Publishers' mission is to publish books that contribute in funda-
mental ways to building an ecologically sustainable and just society, and to do
so with the least possible impact on the environment, in a manner that models
this vision. We are committed to doing this not just through education, but
through action. This book is one step toward ending global deforestation and
climate change. It is printed on Forest Stewardship Council-certified acid-free
paper that is **100% post-consumer recycled** (100% old growth forest-free),
processed chlorine free, and printed with vegetable-based, low-VOC inks, with
covers produced using FSC-certified stock. New Society also works to reduce its
carbon footprint, and purchases carbon offsets based on an annual audit to ensure
a carbon neutral footprint. For further information, or to browse our full list of
books and purchase securely, visit our website at: **www.newsociety.com**

Library and Archives Canada Cataloguing in Publication

Chiras, Daniel D
    Solar electricity basics : a green energy guide / Dan Chiras.

Includes index.

ISBN 978-0-86571-618-6
    1. Solar energy. 2. Photovoltaic power systems. I. Title.

TK1087.C46 2010   621.31'244   C2010-901730-7

NEW SOCIETY PUBLISHERS

Mixed Sources
Cert no. SW-COC-001271
© 1996 FSC
FSC

Contents

Books for Wiser Living
recommended by *Mother Earth News*

Today, more than ever before, our society is seeking ways to live more conscientiously. To help bring you the very best inspiration and information about greener, more sustainable lifestyles, *Mother Earth News* is recommending select New Society Publishers' books to its readers. For more than 30 years, *Mother Earth* has been North America's "Original Guide to Living Wisely," creating books and magazines for people with a passion for self-reliance and a desire to live in harmony with nature. Across the countryside and in our cities, New Society Publishers and *Mother Earth* are leading the way to a wiser, more sustainable world.

CHAPTER 1

AN INTRODUCTION
TO SOLAR ELECTRICITY

Ninety-three million miles from Earth is the source of virtually all energy on Earth, the Sun. It powers plants and the animals that feed on them. It is the source of energy in coal, oil, and natural gas, formed millions of years ago from ancient photosynthetic organisms. Although I barely tap into this immense energy resource, that won't be the case forever. With fossil fuel supplies like oil on the decline, the Sun will become our primary energy source. Most younger readers will see the day when solar and wind energy (which is generated by sunlight) become the dominant source of energy.

In recent years, rising concern over the high social, economic, and environmental costs of global warming has stirred intense interest in renewable energy, especially solar energy. Interest in solar and other forms of renewable energy has also been bolstered by concern over declining fossil fuel resources, especially oil and natural gas. These concerns, in turn, have created favorable policies at the state and national level that are making it more affordable for many businesses and tens of thousands of homeowners to install solar systems of all types.

This book focuses on one particular type of solar system, solar electric systems for homes and businesses. They are also known as *photovoltaic systems* or *PV systems* for short.

An Overview of Solar Systems

As their name implies, solar electric systems convert the Sun's energy into electricity. This conversion takes place in solar modules, also commonly referred to as solar panels. A solar module consists of numerous solar cells (Figure 1.1). Solar cells, in turn, are made from one of the most abundant chemical substances on Earth, silicon — a component of sand and quartz that also makes up much of Earth's crust.

Solar cells are wired together and encased in plastic and glass with a metal frame to create solar modules. The plastic and glass layers protect the solar cells from the elements, especially moisture. Two or more modules are typically mounted on a rack and wired together. Together, the rack and solar modules are referred to as a *solar array*.

Electricity generated by a PV array flows via wires to yet another component of the system, the *inverter*. The inverter converts the direct current electricity produced by solar cells into the alternating-current electricity commonly used in US homes and businesses.

In this book, I'll focus on residential and small business-sized solar electric systems whose rated power typically ranges from 1,000 to 10,000 watts. What is rated power?

Fig. 1.1: *Solar electric systems generally consist of two or more modules wired together to form an array. Each module consists of multiple solar cells wired in series to increase the voltage.*

Rated Power and Capacity

Rated power is the instantaneous output of a solar module measured in watts under standard test conditions (STC). Watts is a measure of the rate of flow of energy. Most of us are familiar with the term "watts." It is used to rate devices that consume electricity, for example, 100-watt light bulbs and 1,200-watt microwaves. This number describes the amount of power a device consumes.

Watts is also used to rate technologies that produce electricity, such as solar modules and wind turbines. You might, for instance, install a 3,000-watt solar system on your home. Three thousand watts is its *rated output*.

The rated output of a PV system is determined by adding up the rated output of the modules in the system. The rated output of a module is determined under standard test conditions. The test takes place at 77°F (25°C) — when the modules are only slightly warm. Light at 1,000 watts per square meter, which is equivalent to full sun, is flashed on the module and the output is measured (in watts). In this test, the light rays arrive perpendicular to the module. (Light rays striking the module perpendicular to the surface result in the greatest absorption of sunlight.)

Rating modules under standard test conditions provides a number that buyers can use when comparing one module to another. It is also used when sizing a solar system.

Most residential solar electric systems fall within the range of 1,000 to 6,000 watts or 1 to 6 kilowatts (kW) (1,000 watts is equal to 1 kilowatt), although PV systems as large as 9,000 or 10,000 watts are sometimes required for very large homes or businesses. It is important to note, however, that the rated output of a PV system doesn't mean the system produces that much electricity at all times. A 3 kW PV system, for instance, won't produce 3,000 watts of electricity all the time the Sun's shining on it. It only produces this amount under standard test conditions.

When mounted outdoors, PV modules typically operate at higher temperatures than the 77°F used for standard test conditions. That's because infrared radiation (heat) in sunlight striking

PV modules warms them up pretty quickly. To achieve a module temperature of 77°F (25°C) under full sun, like that measured in the laboratory, the air temperature must be quite low — about 23° to 32°F (0° to -5°C), not a typical temperature for PV modules in most locations most of the year. To better simulate real-world conditions, the solar industry has developed an alternative rating system under different conditions. They call it PTC, which stands for *PV-USA Test Conditions* (which, in turn, stands for *PV for Utility Scale Applications Test Conditions*).

PTC were developed at the PV-USA test site in Davis, California. In this system, the modules are tested with the ambient temperature at 68°F (20°C). The output is measured when the Sun's irradiance reaches 1,000 watts per square meter. The conditions also take into account the cooling effect of wind. Wind speed in the test is 2.24 miles per hour (or 1 meter per second) 33 feet (10 meters) above the surface of the ground. Although the cell temperature varies with different modules, it is typically about 113°F (45°C) under PTC.

PTC ratings are generally 10% lower than STC ratings, because power output decreases with rising temperature. You'll find the power output falls about 0.5% for every 1°C increase in temperature. When shopping for PV modules, ignore the nameplate rating the manufacturers provide and look up the PTC ratings. Some manufacturers publish these data on their spec sheets; others don't. You can look up PTC ratings on the California Energy Commission's website: consumerenergycenter.org.

Applications

Once a curiosity, solar electric systems are becoming commonplace, even in less-than-sunny areas such as Germany and Japan. (Germany and Japan are the largest markets for PV modules in the world today.) While many solar electric systems are being installed around the world to provide electricity to homes, they are also being installed on schools, small businesses, and office buildings — even skyscrapers like 4 Times Square in New York City, home of the

NASDAQ. Many large corporations such as Microsoft, Toyota, and Google have installed large solar electric systems. Some electric utilities are also installing large PV arrays to generate electricity.

Even some colleges and airports are installing PV systems. On a smaller scale, ranchers often install solar electric systems to power electric fences to contain their livestock or to pump water into remote stock-watering tanks.

Solar electricity is proving to be extremely useful on sailboats. Their small PV systems often power lights, fans, radio communications, GPS systems, and refrigerators. Many recreational vehicles (RVs) are also equipped with small PV systems to power microwaves, TVs, and satellite receivers.

Many bus stops and parking lots are illuminated by solar electricity, as are portable information signs used at construction sites. Numerous police departments now haul solar-powered radar units to neighborhoods to discourage speeding. These units display a car's speed and warn drivers if they're exceeding the speed limit. Backpackers and river runners can even take roll-up solar modules with them on their ventures into the wild to power electronics (Figure 1.2).

Solar electric systems are well suited for remote applications where it is too costly to run power lines. Emergency call boxes

Fig. 1.2: *This small solar device is designed to charge portable electronic equipment, including cell phones, PDAs, and other small devices such as IPods.*

Power and Energy: What's the Difference?

The terms "power" and "energy" are frequently used, but often misunderstood. In the electrical world, power is measured in watts or kilowatts. Power is an instantaneous measure, like the speed of a car. In the electrical world, power (watts) is a measure of the *rate of flow of energy*. Engineers and scientists rate electrical loads, which are devices that consume electricity such as light bulbs and electric motors, in watts or kilowatts. For example, an electric toaster might be rated at 1,000 watts. Scientists and engineers also use watts, as noted in the text, to rate electric-generating technologies such as PV systems. A solar system, for instance, might be rated at 5,000 watts. Since 1,000 watts is 1 kilowatt, a 5,000-watt system is also referred to as a 5-kilowatt system or a 5-kW system.

Energy, in contrast, is *power consumption* or *production* over time. A light bulb that consumes 100 watts for one hour, consumes 100 watt-hours of electricity. If it operates for 10 hours, it uses 1,000-watt-hours of energy or 1 kilowatt-hour of energy. Both watt-hours and kilowatt-hours are measures of energy production and use. A solar electric system producing electricity *at a rate of* 1,000 watts for four hour produces 4,000 watt-hours, or four kilowatt-hours of energy.

found in many remote stretches of highway in North America are powered by solar electricity, as are many highway warning lights (Figure 1.3). Solar electricity is also used to power remote monitoring stations that collect data on rainfall, temperature, and seismic activity. Stream flow monitors on many US rivers and streams rely on solar-powered transmitters to beam data to solar-powered satellites. The data is then beamed back to Earth to the US Geological Survey, where it is processed and disseminated. PVs also allow scientists to gather and transmit data back to their labs from remote sites, like the tropical rain forests of Central and South America.

Fig. 1.3: *Solar modules are the product of choice in remote locations. Solar-powered emergency call boxes like the one shown here on a remote highway in northern California are now found along rural highways in many states.*

DAN CHIRAS

Solar electric modules often power lights on buoys, which are vital for nighttime navigation on large rivers like the Saint Lawrence Seaway. Railroad signals and aircraft warning beacons are also often solar powered.

PV modules are used to boost radio, television, and telephone signals. Signals from these sources are often transmitted over long distances. For successful transmission, however, they must be periodically amplified at relay towers. The towers are often situated in inaccessible locations, far from power lines. Because they are reliable and require little, if any, maintenance, PV systems are ideal for such applications. They make it possible for us to communicate across long distances. Next time you make a long-distance telephone call from a phone on a landline, rest assured solar energy is making it possible.

While PV systems are becoming very popular in more developed countries, they're also widely used in the developing world. They are, for instance, being installed in remote villages in less developed countries to power lights and televisions. They're often installed to power refrigerators and freezers to store medicines. PV systems may also be used to power pumps to provide water for villages.

The ultimate in remote and mobile applications, however, has to be the satellite. Virtually all military and telecommunications satellites are powered by solar electricity, as is the International Space Station.

World Solar Energy Resources

Although solar electricity is growing in popularity, it provides only a tiny fraction of today's electrical demand. However, as global supplies of fossil fuel resources decline and as concerns over global climate change increase, solar electric systems could become a major source of electricity, — along with wind systems and other renewable energy technologies. But is there enough solar energy to meet our needs?

Although solar energy is not evenly distributed over the Earth's surface, significant resources are found on every continent. "Solar energy's potential is off the chart," write energy experts Ken Zweibel, James Mason, and Vasilis Fthenakis in a December 2007 article, "A Solar Grand Plan," published in *Scientific American* magazine. The solar energy striking the Earth in a 40-minute period is equal to all the energy human society consumes in a year. In the not-too-distant future, solar electric systems mounted on our homes and businesses or giant commercial solar systems could produce a substantial portion of the electricity to meet the needs of homes, businesses, farms, ranches, schools, and factories.

Another solar technology currently used by a handful of utilities is known as *solar thermal electricity*. Solar thermal electric systems concentrate sunlight energy to generate heat that's used to boil water. Steam generated from this process is used to spin a turbine connected to a generator that makes electricity (Figure 1.4). Some of the newest solar thermal electric systems even store hot water so electricity can be generated on cloudy days or at night.

Despite what critics say, solar energy is an abundant resource and many forms of solar energy-capture technologies are affordable right now. Solar energy will very likely play a major role in our energy future, along with wind and other renewables. They have to. Fossil fuels like oil are finite. In fact, conventional oil (crude oil, not shale oil) could be economically depleted within 30 to 50 years.

a

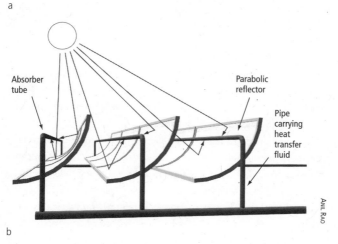

Absorber tube

Parabolic reflector

Pipe carrying heat transfer fluid

ANIL RAO

SANDIA NATIONAL LABORATORY

b

Fig. 1.4: (a) Solar thermal electric systems like the one shown here produce electricity at cost-competitive rates. (b) Sunlight is focused on a pipe that carries a heat transfer fluid. The heat is transferred to water at a central point, causing it to boil. Steam produced by the boiling water drives a turbine. The spinning turbine is attached to a generator that produces electricity.

Many energy experts believe that the production of crude oil has peaked and is on the decline. (This is partly responsible for the record high gasoline prices in 2007 and much of 2008.) Some energy experts believe that global natural gas production could peak between 2015 and 2025. Coal, which is abundant, is still a finite resource and, lest we forget, a major source of the greenhouse gas carbon dioxide, the main cause of global warming.

The Sun, on the other hand, will continue to shine for at least 5 billion more years. With 30 to 50 years of oil left and 5 billion years of sunlight, what's the future going to look like?

The Pros and Cons of Solar Electricity

Solar energy is a seemingly ideal fuel source. It's clean. It's free. It's abundant, and its use could ease many of the world's most pressing environmental problems, chief among them global climate change. Because solar energy has its share of critics, it's worth taking a look at both the pros and cons of solar energy and responding to the criticisms.

Availability and Variability

Although the Sun shines 24 hours a day and beams down on the Earth at all times, half the planet is always immersed in darkness. This poses a problem, because modern societies consume electricity 24 hours a day, 365 days a year.

Another problem is the daily variability of solar energy. That is, even during daylight hours clouds can block the Sun, sometimes for days on end. If PV systems are unable to generate electricity 24 hours a day like coal-fired and nuclear power plants, how can we use them to power our 24-hour-per-day demand for electricity?

Homeowners like myself who live off-grid (not connected to the electrical grid) solve the problem by installing batteries to store electricity to meet their nighttime demand and to supply electricity for use on cloudy days. As a result, they are supplied with electricity 24 hours a day, 365 days a year by PV systems.

The Sun's variable nature can also be offset by coupling solar electric systems with other renewable energy sources, for example, wind-electric systems, or micro hydro systems. Wind systems, for instance, generate electricity day and night — so long as the winds blow. Micro hydro systems tap the energy of flowing water in streams or rivers. Either one can be used to generate electricity to supplement a PV system, compensating for the Sun's natural variability. Solar and wind are especially good partners. Figure 1.5 shows data on the solar and wind resources at my renewable energy education center in eastern Missouri. As illustrated, the Sun shines a lot in the spring, summer, and early fall but less so during the winter. During winter, however, the winds blow more often and much more forcefully. A wind turbine could easily make up for the reduced output of a PV system, ensuring a reliable, year-round supply of electricity at this and other similar sites.

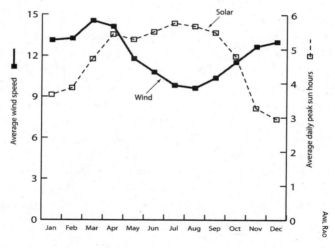

Fig. 1.5: *This graph of solar and wind energy resources at The Evergreen Institute's Center for Renewable Energy and Green Building in east central Missouri (operated by Dan Chiras) illustrates how complementary wind and sun are. As illustrated, the sunlight is fairly plentiful in the spring, summer, and fall, but not the winter. Wind picks up in the fall, winter, and early spring. A hybrid solar-electric/wind system can tap into these resources, providing an abundance of electricity year round.*

While the Sun's variability can easily be overcome on the individual level, can society find a way to meet its 24-hour-per-day needs for electrical energy from the Sun? Scientists and engineers are currently developing numerous ingenious technologies to store solar electricity. Batteries are not high on the list, however. Why?

To store massive amounts of electricity to power factories, stores, and homes, we'd need equally massive battery banks. Because they would be costly and would require huge investments, scientists are seeking a variety of other, potentially more cost-effective options. One option is the solar thermal electric system mentioned earlier, which stores surplus solar-heated water to run electric generators at night or during cloudy periods.

Another option is the use of solar electricity to power air compressors. They'd produce compressed air that could be stored in abandoned underground mines. When electricity is needed, the compressed air would be released through a turbine, not unlike those found in conventional power plants. The blades of the turbine would be attached to a shaft that is attached to a generator that produces electricity.

Surplus solar electricity could also be used to generate hydrogen gas from water. Hydrogen gas is created when electricity is run through water. This process, known as electrolysis, splits the water into its components, hydrogen and oxygen, both gases. Hydrogen can be stored in tanks and later burned to produce hot air or to heat water to produce steam. Hot air and steam can be used to spin a turbine attached to a generator. Hydrogen could also be fed into a fuel cell, which produces electricity.

As in residential systems, electricity can also be supplied on cloudy days or at night by commercial wind farms (Figure 1.6). They could provide power to supplement solar systems because the winds often blow when the Sun is behind clouds (during storms, for instance). Other renewable energy technologies can also be used to complement solar energy. Hydroelectric plants and biomass facilities, for instance, could be used to ensure a continuous supply of renewable energy in a system that's finely tuned to switch from one energy source to another. In Canada, hydroelectric facilities can be turned on and off as needed to meet demand. Such systems could be used to produce electricity when demand exceeds the capacity of commercial solar systems or at night.

Shortfalls could also be offset on a local or regional level by transferring electricity from areas of surplus solar and/or wind production to areas of insufficient electrical production. Surplus solar-generated electricity from Colorado, for example, could be shipped via the electrical grid to neighboring Wyoming, New Mexico, and Nebraska when needed — or vice versa.

Solar's variability can also be offset by natural gas-fired power plants and newer coal-fired plants that burn pulverized coal. Both can be started or stopped, or throttled up or down, to provide additional electricity. These facilities could serve as an excellent backup source as we transition to a renewable energy future.

With smart planning and careful design, we can meet a good portion of our electrical needs from this seemingly capricious resource.

Fig. 1.6: *My boys, Skyler and Forrest, check out a large wind turbine at a wind farm in Canastota, NY. Wind farms like this one are popping up across the nation, indeed across the world, producing clean, renewable electricity to power our future.*

Dan Chiras

Aesthetics

While many of us view a solar electric array as a thing of great value, our neighbors don't always share our opinions. Some neighborhood associations have banned PV systems.

Ironically, those who object to solar electric systems rarely complain about the visual blight in our environs, among them cell phone towers, water towers, electric transmission lines, and billboards. One reason that these common eyesores draw little attention is that they have resided in our communities for decades. We've grown used to ubiquitous electric lines and radio towers.

Fortunately there are ways to mount a solar array so that it blends seamlessly with a roof. As you'll learn in Chapter 8, solar modules can be flush mounted on roofs (Figure 1.7). There's also a solar product that can be applied directly to a certain type of metal roof, creating an even lower-profile array (Figure 1.8). Solar arrays can also be mounted on a pole or rack anchored to the ground that can be placed in sunny backyards — out of a neighbor's line of sight.

Cost

Perhaps the biggest disadvantage of solar electric systems is that they're costly — very costly. Although the cost of solar cells has

Fig. 1.7: *Solar arrays can be flush mounted, that is, mounted close to the roof to reduce their visibility. The arrays are typically mounted on racks three to six inches from the surface of the roof which helps cool the modules. While aesthetically appealing,*

flush-mounted arrays often produce less electricity than a rack-mounted array because the output of PV modules typically decreases at higher temperatures.

Fig. 1.8: *Uni-Solar produces a thin solar laminate that can be applied directly to standing seam metal roofs.*

ENERGY CONVERSION DEVICES/UNI-SOLAR

fallen precipitously, from around $50 per watt in the mid 1970s to $5 per watt in the early 2000s then to about $3.50 a watt in 2009, solar electric systems are one of the most expensive means of generating electricity — but only if you ignore the environmental costs of conventional power and the generous subsidies these technologies receive from taxpayers.

Although solar electric systems are expensive, there are ways to lower the cost — often substantially. And there are factors that make a system competitive with conventional electricity. If, for instance, you live in an area with lots of sunshine and high electrical rates, such as southern California or Hawaii, a PV system competes very well with conventional electricity. Even in areas with low sunshine but high electrical rates, such as Germany, PV is economically competitive. Financial incentives for PV systems from local utilities or the state and federal government drive costs down, often making solar electricity cost competitive with conventional sources.

If you are building a home more than a few tenths of a mile from a power line, solar electricity can also compete with utility power. That's because utility companies often charge customers a large fee to connect to the utility grid. You could, for instance, pay $20,000 to connect to the electric line, even if you're only a few tenths of mile away from a power line. Line extension fees don't pay for a single kilowatt-hour of electricity; they only cover the cost of the transformer, poles, wires, electrical meter, and installation. You'll pay the cost of the connection either up front or pro-rated over many years.

The Advantages of Solar Electric Systems

Although solar electricity, like any fuel, has some downsides, they're clearly not insurmountable and, many believe, they are outweighed by their advantages. One of the most important advantages is that solar energy is an abundant, renewable resource. While natural gas, oil, coal, and nuclear fuels are finite and on the decline, solar energy will be available to us as long as the Sun continues to shine — for at least 5 billion years.

Solar energy is a clean energy resource, too. By reducing our reliance on coal-fired power plants, solar electricity could help homeowners and businesses reduce their contribution to a host of environmental problems, among them acid rain, global climate change, habitat destruction, and species extinction. Solar electricity could even replace costly, environmentally risky nuclear power plants. Nuclear power plants cost upwards of $6 to $ billion, and no long term solution has been enacted to store the high-level radioactive waste they produce.

Solar energy could help us decrease our reliance on declining and costly supplies of fossil fuels like natural gas. Solar could also help us decrease our reliance on oil. Although very little electricity in the United States comes from oil, electricity generated by solar electric systems could be used to power electric or plug-in hybrid cars and trucks, reducing our demand for gasoline and diesel fuel, both of which come from oil (Figure 1.9). And, although the production of solar electric systems does have its impacts, all in all it is a relatively benign technology compared to fossil fuel and nuclear power plants.

Another benefit of solar electricity is that, unlike oil, coal, and nuclear energy, the fuel is free. Moreover, solar energy is not owned or controlled by hostile foreign states or one of the dozen or so

Fig. 1.9: *Plug-in hybrids like the one shown here and electric cars (not shown) will very likely play a huge role in personal transportation in the future. Electric cars with longer-range battery banks could be used for commuting and for short trips, (under 200 miles) while plug-in hybrids could be used for commuting and long-distance trips.*

influential energy companies that largely dictate energy policy, especially in the United States. Because the fuel is free and will remain so, solar energy can provide a hedge against inflation, caused in part by ever-increasing fuel costs.

An increasing reliance on solar and wind energy could also ease political tensions worldwide. Solar and other renewable energy resources could alleviate the need for costly military operations aimed at stabilizing the Mideast, a region where the largest oil reserves reside. Because the Sun is not owned or controlled by the Middle East, we'll never fight a war over solar or other renewable energy resources. Not a drop of human blood will be shed to ensure a steady supply of solar energy to fuel our economy.

Yet another advantage of solar-generated electricity is that it uses existing infrastructure — the electrical grid — and technologies in use today such as electric toasters, microwaves, and the like. A transition to solar electricity could occur fairly seamlessly.

Solar electricity is also modular. You can add on to a system over time. If you can only afford a small system, you can start small and expand your system as money becomes available.

Solar electricity could provide substantial economic benefits for local, state, and regional economies. And solar electricity does not require extensive use of water, an increasing problem for coal, nuclear, and gas-fired power plants, particularly in the western United States and in arid regions.

Purpose of this Book

As noted earlier, this book focuses on small solar electric systems suitable for homes and small businesses. I have strived to explain facts and concepts clearly and accurately, introducing key terms and concepts as needed, and repeating them as necessary. This book is written for individuals who aren't necessarily well versed in electricity or electronics.

When you finish reading and studying the material in this book, you'll know an amazing amount about solar energy and solar electric energy systems. You will have the knowledge required to assess your electrical consumption and the solar resource at your site. You will also be able to determine if a solar electric system will meet your needs and if it makes economic sense. You will have a good working knowledge of the key components of solar electric systems, especially PV modules, racks, controllers, batteries, inverters, and backup generators. This book will also help you know what to look for when shopping for a PV system. You'll also know how PV systems are installed.

It is important to note that this book is not an installation manual. When you're done reading, you won't be qualified to install a solar electric system — that usually requires a few hands-on workshops like those I teach at The Evergreen Institute's Center for Renewable Energy and Green Building (www.evergreeninstitute.org). Even so, this book is a good start. You'll understand much of what's needed to install a system or even launch a new career in PVs. If you choose to hire a solar energy professional to install a system (a route I highly recommend) you'll be thankful you've read and studied the material in this book. The more you know, the more informed input you will have into your system design, components, siting, and installation — and the more likely you'll be happy with your purchase.

This book should also help you develop realistic expectations. I believe that those interested in installing renewable energy systems need to proceed with eyes wide open. Knowing the shortcomings and pitfalls of solar electric systems (or any renewable energy technology, for that matter) helps us avoid mistakes and prevents the disappointments that often result from unrealistic expectations.

Organization of this Book

We'll begin in the next chapter by studying the Sun and the energy it produces. We will discuss important terms and concepts. You'll learn about *altitude angle*, *azimuth angle*, and the all-important *tilt angle*.

In Chapter 3, we'll explore solar electricity — the history of PVs, the types of solar cells on the market today, and how solar cells generate electricity.

In Chapter 4, you will learn how to assess your electrical energy needs and how to determine if your site has enough solar energy to meet them. You'll also learn why energy-efficiency measures are so important and how they can reduce your initial system cost.

In Chapter 5, we'll examine three types of residential solar electric systems and the components of each one. We'll also examine hybrid systems — for example, wind-PV systems — and how they complement each other, providing a reliable, year-round supply of electricity. You'll learn about net metering and ways to make solar electricity affordable.

Chapter 6 introduces you to inverters, vital components of solar electric systems. You'll learn what they do and how they operate. We'll also provide some shopping tips — ideas on what you should look for when buying an inverter.

In Chapter 7, we'll tackle batteries and controllers, key components of battery-based PV systems. You will learn about the types of batteries you can install, how to care for them, and ways to reduce battery maintenance. I will point out common mistakes people make with their batteries and ways to avoid making those same mistakes. You will also learn about battery safety. We'll finish with a discussion of charge controllers and backup generators.

In Chapter 8, I'll provide an overview of solar electric system installation and maintenance. You'll learn how to size a PV system and about the various mounting options.

In Chapter 9, we'll explore a range of issues such as permits, covenants, and utility interconnection. We'll discuss whether you should install a system yourself or hire a professional and, if you choose the latter, how to locate a competent installer.

CHAPTER 2

UNDERSTANDING THE SUN AND SOLAR ENERGY

The Sun lies in the center of our solar system, approximately 93 million miles from Earth. Composed primarily of hydrogen and small amounts of helium, the Sun is a massive fusion reactor. In the Sun's core intense pressure and heat force hydrogen atoms to unite, or fuse, creating slightly larger helium atoms. In this process, immense amounts of energy are released; this energy migrates to the surface of the Sun, and then radiates out into space, primarily as light and heat.

Solar radiation streaming into space strikes the Earth, warming and lighting our planet and fueling aquatic and terrestrial ecosystems. According to French energy expert Jean-Marc Jancovici, "the solar energy received each year by the Earth is roughly … 10,000 times the total energy consumed by humanity." To replace *all* the oil, coal, gas, and uranium currently used to power human society with solar energy, we'd need to capture a mere 0.01% of the energy of the sunlight striking the Earth each day. According to the US Department of Energy's National Renewable Energy Laboratory (NREL), to generate the electricity the United States consumes, we'd only need to install PVs on 7% of the total land surface area currently occupied by cities and homes. We could achieve this by installing PVs on rooftops, over parking lots, and on the sides of buildings. We wouldn't need to appropriate a single acre of new land to make PV our primary energy source!

In this chapter, I'll explore the Sun and solar radiation and key concepts vital to solar electric systems.

Understanding Solar Radiation

The Sun's output is known as *solar radiation*. As shown in Figure 2.1, solar radiation ranges from high-energy, short-wavelength gamma rays to low-energy, long-wave radiation known as radio waves. In between these extremes, starting from the short-wave end of the spectrum are x-rays, ultraviolet radiation, visible light, and heat (infrared radiation).

While the Sun releases numerous forms of energy, most of it (about 40%) is infrared radiation (heat) and visible light (about 55%). Traveling at a speed of 186,000 miles per second, solar energy takes 8.3 minutes to make its 93-million-mile journey from the Sun to the Earth. Most PV modules capture the energy contained in the visible and lower end of the infrared portions of the spectrum.

Solar radiation travels virtually unimpeded through space until it encounters the Earth's atmosphere. In the outer portion of the atmosphere (a region known as the stratosphere) ozone molecules

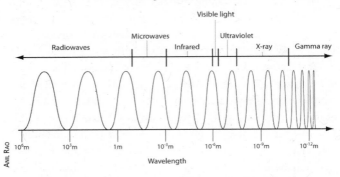

Fig. 2.1: *Electromagnetic Spectrum. The sun produces a wide range of electromagnetic radiation from gamma rays to radiowaves, shown here. PV cells convert visible light into electric energy, specifically, direct current electricity.*

(O_3) in the ozone layer absorb much (99%) of the incoming *ultraviolet* radiation, dramatically reducing our exposure to this potentially harmful form of solar radiation. As sunlight passes through the lower portion of the atmosphere (the troposphere), it encounters clouds, water vapor, and dust. These may absorb the Sun's rays or reflect them back into space, reducing the amount of sunlight striking the Earth's surface.

Irradiance

The amount of solar radiation striking a square meter of the Earth's atmosphere or the Earth's surface is known as *irradiance*. It is measured in watts per square meter (W/m^2). Solar irradiance measured just before the Sun's radiation enters the Earth's atmosphere is about 1,366 W/m^2. On a clear day, nearly 30% of the Sun's radiant energy is absorbed and converted into heat or reflected by dust and water vapor in the Earth's atmosphere. By the time the incoming solar radiation reaches a solar array on a roof, the incoming solar radiation is reduced to about 1,000 W/m^2.

Solar irradiance varies during daylight hours at any given site. At night, solar irradiance is zero. As the Sun rises, irradiance increases, peaking around noon. From noon until sunset, irradiance slowly decreases, falling once again to zero at night. These changes in irradiance are determined by the angle of the Sun's rays, which changes continuously as the Earth rotates on its axis. The angle at which the Sun's rays strike the Earth affects both the *energy density*, described shortly (Figure 2.2), and the amount of atmosphere through which sunlight must travel to reach the Earth's surface (Figure 2.3). Let's start with energy density.

As shown in Figure 2.2, low-angled sunlight delivers much less energy per square meter than high-angled sunlight. The result is a decrease in energy density. The lower the density, the lower the irradiance. Early in the morning, then, irradiance is low. As the Sun makes its way across the sky, however, irradiance increases.

Irradiance is also influenced by the amount of atmosphere through which the sunlight passes, as shown in Figure 2.3. The

Fig. 2.2:
Energy Density. Surfaces perpendicular to the incoming rays absorb more solar energy than surfaces not perpendicular, as illustrated here.

Fig. 2.3:
Atmospheric Air Mass and Irradiance. Early and late in the day, sunlight travels through more air in the Earth's atmosphere, which decreases the amount of energy reaching a solar electric array, decreasing its

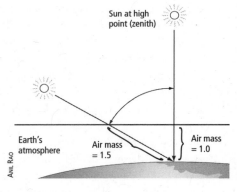

output. Maximum output occurs when the sun's rays pass through the least amount of atmosphere, at solar noon. Three-quarters of the daily output from a solar array occurs between 9 AM to 3 PM.

more atmosphere through which sunlight passes, the more filtering occurs. The more filtering, the less sunlight makes it to Earth, and the lower the irradiance.

Irradiation

Irradiance is an important measurement, but what most solar installers need to know is irradiance over time — the amount of energy they can expect to strike a PV array. Irradiance over a period of time is referred to as *solar irradiation*. It's expressed as watts per square meter striking the Earth's surface (or a PV module) for some specified period of time — usually an hour or a day. The units of hourly irradiation are expressed as watt-hours per square meter. For example, solar radiation of 500 watts of solar energy striking a square meter for an hour is 500 watt-hours per square meter. Solar radiation of 1,000 watts per square meter over two hours is 2,000 watt-hours per square meter.

To help keep irradiance and irradiation straight, you can think of irradiance as a measure of instantaneous power (power is measured in watts as noted in Chapter 1). Irradiation, on the other hand, is a measure of power over some period of time and is, therefore, a measure of energy. Teachers help students keep the terms straight by likening irradiance to the speed of a car. Like irradiance, speed is an instantaneous measurement. It simply tells us how fast a car is moving. It's a rate. Irradiation is akin to the distance a vehicle travels. Distance, of course, is determined by multiplying the speed of a vehicle by the time it travels at a given speed. In a car, the faster you travel, the greater the distance you'll cover in a given time. In solar energy, the greater the irradiance over a period of time, the greater the solar irradiation.

Figure 2.4 illustrates the concepts graphically. In this diagram, irradiance is the single black line in the graph — the number of

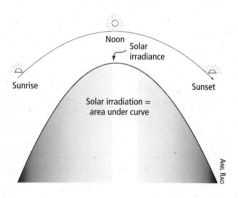

Fig. 2.4: *Solar Irradiance and Solar Irradiation. This graph shows both solar irradiance (watts per square meter striking the Earth's surface) and solar irradiation (watts per square meter per day). Note that irradiation is total solar irradiance over some period of time, usually a day. Solar irradiation is the area under the curve.*

watts per square meter at any moment in time. The area under the curve is solar irradiation, the total solar irradiance during a given period — in this graph, it's the irradiance occurring in a day.

Knowing an area's solar irradiation is useful to professional installers and do-it-yourselfers when sizing a system. Solar irradiation measurements are also useful to utilities that base rebates on the projected electrical production of their customers' PV systems.

Peak Sun, Peak Sun Hours, and Insolation

Another measurement installers use when designing PV systems is *peak sun*. Peak sun is the maximum solar irradiance available at most locations on the Earth's surface on a clear day — 1,000 W/m². One hour of peak sun is known as a *peak sun hour*. Figure 2.5 shows a map that shows the daily average peak sun hours in the United States and Canada measured as kW/m²/day.

Peak sun hours is a measure of solar irradiation (watts per m² per day). It is also commonly referred to as *solar insolation* or simply *insolation*. Installers determine solar insolation by consulting detailed state maps or tables. They use this number to determine how large an array must be to meet a customer's needs. It's important to note that the average peak sun hours per day for a given location doesn't mean that the Sun shines at peak intensity during that entire period. In fact, peak sun conditions — solar irradiance equal to 1,000 Watts/m² — will very likely only occur one or two hours a day. So how do scientists calculate peak sun hours per day?

Peak sun hours are calculated at a given location by determining the total irradiation (watt-hours per square meter) received during daylight hours and dividing that number by 1,000 watts per square meter. On a summer day, for instance, the solar irradiance may average 600 watts over 12 hours. As a result, the total solar irradiation is 7,200 watt-hours, and peak sun hours is 7,200 watt-hours divided by 1,000 or 7.2 hours of peak sun. Even though the peak sun hours per day is 7.2, solar irradiance may only reach 1,000 watts/m² for two hours.

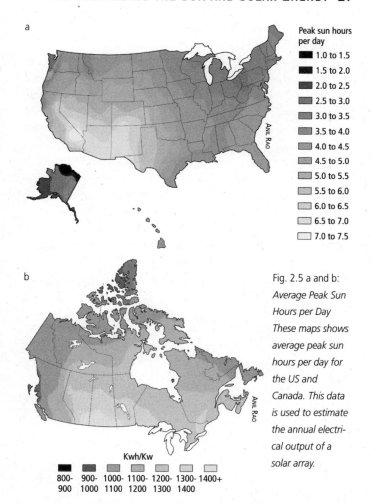

Fig. 2.5 a and b:
Average Peak Sun Hours per Day These maps shows average peak sun hours per day for the US and Canada. This data is used to estimate the annual electrical output of a solar array.

The Sun and the Earth: Understanding the Relationships

Now that you understand irradiance, irradiation, peak sun, peak sun hours and isolation, let's examine the geometric relationships between the Earth and Sun.

Day Length and Altitude Angle:
The Earth's Tilt and Orbit Around the Sun

As you learned in grade school, the Earth orbits around the Sun, completing its path every 365 days. As shown in Figure 2.6, the Earth's axis is tilted 23.5°. The Earth maintains this angle throughout the year as it orbits around the Sun. Look carefully at Figure 2.7 to see that the angle remains fixed — almost as if the Earth were attached to a wire anchored to a fixed point in outer space. Because the Earth's tilt remains constant, the Northern Hemisphere is tilted away from the Sun during the winter. As a result, the Sun's rays enter and pass through Earth's atmosphere at a very low angle. Sunlight penetrating at a low angle passes through more atmosphere and is therefore absorbed or scattered by more dust and water vapor, as shown in Figure 2.3. This, in turn, reduces irradiance, which reduces the output of a solar array.

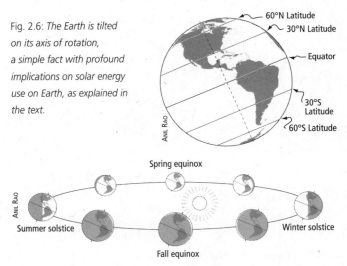

Fig. 2.6: *The Earth is tilted on its axis of rotation, a simple fact with profound implications on solar energy use on Earth, as explained in the text.*

— 60°N Latitude
— 30°N Latitude
— Equator
30°S Latitude
60°S Latitude

Spring equinox

Summer solstice

Winter solstice

Fall equinox

Fig. 2.7: *As any school child can tell you, the Earth orbits around the Sun. Its orbit is not circular, but rather elliptical, as shown here. Notice that the Earth is tilted on its axis and is farthest from the sun during the summer. Because the northern hemisphere is angled toward the sun, however, our summers are warm.*

Irradiance is also lowered because the density of sunlight striking the Earth's surface is reduced when it strikes a surface at an angle (described earlier). As shown in Figure 2.2, a surface perpendicular to the Sun's rays absorbs more solar energy than one that's tilted away from it. As a result, low-angled sunlight delivers much less energy per square meter of surface in the winter than it does during summer.

Solar gain is also reduced in the winter because days are shorter — that is, there are fewer hours of daylight during winter months. Day length is determined by the angle of the Earth in relation to the Sun. During the winter in the Northern Hemisphere, most of the Sun's rays fall on the Southern Hemisphere. All three factors — lower energy density, increased absorption, and shorter days — reduce the amount of solar energy available to a PV system in the winter.

In the summer in the Northern Hemisphere the Earth is tilted *toward* the Sun, as shown in Figures 2.7 and 2.8. This results in several key changes. One of them is that the Sun is positioned higher in the sky. As a result, sunlight streaming onto the Northern Hemisphere passes through less atmosphere, which reduces absorption and scattering. This, in turn, increases solar irradiance, which increases the output of a PV array. Because a surface perpendicular to the Sun's rays absorbs more solar energy than one that's tilted away from it, the Earth's surface intercepts more energy during the summer as well. Put another way, the high-angled Sun delivers much more energy per square meter of surface area than in the winter. Moreover, days are also longer in the summer. All these factors increase the output of an array.

Figure 2.9 shows the position of the Sun as it "moves" across the sky during different times of the year as a result of the changing relationship between the Earth and the Sun. As just discussed, the Sun "carves" a high path across the summer sky. It reaches its highest arc on June 21, the longest day of the year, also known as the summer solstice. Figure 2.9 also shows that the lowest arc occurs on December 21, the shortest day of the year. This is the winter solstice.

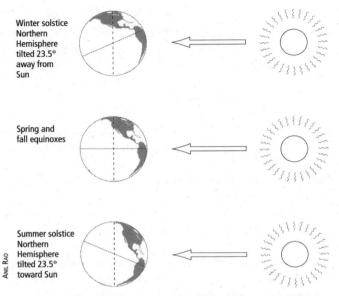

Fig. 2.8: *Summer and Winter Solstice. Notice that the Northern Hemisphere is bathed in sunlight during the summer solstice because the Earth is tilted toward the Sun. The Northern Hemisphere is tilted away from the Sun, during the winter.*

The angle between the Sun and the horizon at any time during the day is referred to as the *altitude angle*.

As shown in Figure 2.11, the altitude angle decreases from the summer solstice to the winter solstice. After the winter solstice, however, the altitude angle increases, growing a little each day, until the summer solstice returns. Day length changes along with altitude angle, decreasing for six months from the summer solstice to the winter solstice, then increasing until the summer solstice arrives once again.

The midpoints in the six-month cycles between the summer and winter solstices are known as *equinoxes*. The word *equinox* is derived from the Latin words *aequus* (equal) and *nox* (night). On the equinoxes, the hours of daylight are nearly equal to the hours of darkness. The spring equinox occurs around March 20 and the fall equinox occurs around September 22. These dates mark the beginning of spring and fall, respectively.

The altitude angle of the Sun is determined seasonally by the angle of the Earth in relation to the Sun, as just noted. The altitude angle is also determined daily by the rotation of the Earth on its axis. As seen in Figure 2.9, the altitude angle increases between sunrise and noon, then decreases to zero once again at sunset.

The Sun's position in the sky relative to a fixed point, such as a PV array, also changes by the minute. Scientists locate the Sun's

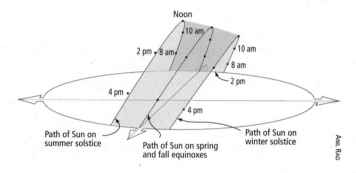

Fig. 2.9: *Solar Path. This drawing shows the position of the Sun in the sky during the day on the summer and winter solstices and the spring and fall equinoxes. This plot shows the solar window — the area you want to keep unshaded so the Sun is available for generating solar electricity.*

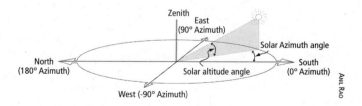

Fig. 2.10: *Altitude and Azimuth Angles. The altitude angle is the angle of the Sun from the horizon. It changes minute by minute as the Earth rotates. It also changes by day. The azimuth angle is the angle of the Sun from true south.*

position in the sky in relation to a fixed point by the *azimuth angle*. As illustrated in Figure 2.10, true south is assigned a value of 0°. East is +90° and west is -90°. North is 180°. The angle between the Sun and 0° south (the reference point) is known as the *solar azimuth angle*. If the Sun is east of south, the azimuth angle falls in the range of 0 to +180°; if it is west of south, it falls between 0 and -180°. Like altitude angle, azimuth angle changes as a result of the Earth's rotation on its axis.

Implications of Sun-Earth Relationships on Solar Installations

Solar modules can be mounted on four types of racks: fixed, seasonally adjustable fixed, single-axis tracker, or a dual-axis tracker.

Fixed racks are oriented to the south and are set at a fixed tilt angle, usually equal to the latitude of the site, give or take a little. As shown in Figure 2.11, the tilt angle is the angle between the surface of the array and an imaginary horizontal line extending back from the bottom of the array. Fixed racks are mounted on the roofs of buildings or on the ground or on poles.

Seasonally adjustable fixed racks resemble fixed racks, but can be adjusted to increase or decrease the tilt angle during various seasons. Tilt angle may be increased in the winter to capture more energy from the low-angled winter sun and decreased in the summer to capture more of the high-angled summer sun.

Single-axis trackers are designed to follow the Sun from sunrise to sunset by automatically adjusting for changes in the azimuth angle of the Sun. Single-axis trackers can increase the output of an array up to 30%.

Dual-axis trackers automatically adjust both the tilt angle and the azimuth angle and can increase the output of an array up to 40%, depending on the location.

For optimum solar gain, solar arrays should be pointed toward true south (in the Northern Hemisphere), not magnetic south. What's the difference?

True north and south are measurements used by surveyors to determine property lines. They are imaginary lines that run parallel

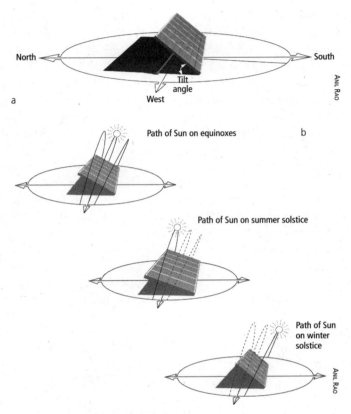

Fig. 2.11 a and b: *Tilt Angle. (a) The tilt angle of an array, shown here, is adjusted according to the altitude angle of the Sun. The tilt angle can be set at one angle year round, known as the optimum angle, or (b) adjusted seasonally or even monthly, although most homeowners find this too much trouble.*

to the lines of longitude, which, of course, run from the North Pole to the South Pole. (True north and south are also known as true geographic north and south.) Magnetic north and south, on the other hand, are determined by the Earth's magnetic field. They are measured by compasses. Unfortunately, magnetic north and south rarely line up with the lines of longitude — that is, they rarely run true north and south. In some areas, magnetic north and south can deviate quite significantly from true north and south. How far magnetic north and south deviate from true north and south is known as the *magnetic declination*.

Figure 2.12 shows the deviation of magnetic north and south — the magnetic declination — from true north and south in North America. You may want to take a moment to study the map. Start by locating your state and then reading the value of the closest isobar. If you live in the eastern United States, you'll notice that the lines are labeled with a minus sign. This indicates a westerly declination — meaning that true south is located west of magnetic south. If you live in the midwestern and western United States, the lines are positive, which indicates an easterly declination. That is, true north and south lie east of magnetic north and south.

To determine the magnetic declination of your home or business, you can consult this map, although a local surveyor or nearby airport can provide a more precise reading. Be sure to ask whether the magnetic declination is east or west. Bear in mind that magnetic

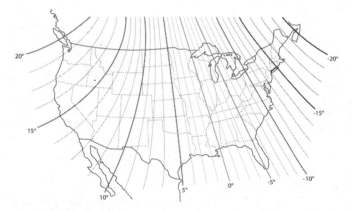

Fig. 2.12: *Magnetic declination. The isobars on this map indicate magnetic declination in the United States. Note that negative numbers indicate a westerly declination (meaning true south lies west of magnetic south). Positive numbers indicate a easterly declination (meaning true south is east of magnetic south).*

north and south not only deviate from true north and south, they change very slightly from one year to the next. Surveyors keep track of the annual variation. Also bear in mind that compass readings at any one site may be slightly off. If you take a magnetic reading too close to your vehicle, for instance, the compass may not point precisely to magnetic south.

As I will point out in subsequent chapters, orienting a solar array to true south is ideal, but it's not always possible. Don't sweat it. You have some leeway. In fact, orienting an array slightly off true south results in only a very slight decrease in output until the azimuth angle is off by about 25°.

Ideally, a solar array should point directly at the Sun from sunrise to sunset to produce the most energy. Doing so optimizes the energy density falling on the array, which was discussed earlier in the chapter. This is the primary benefit of pointing an array at the Sun. Doing so also ensures sunlight strikes the array head on — that is, perpendicular to the surface of the array — at all times. This minimizes reflection of sunlight off the surface of the module and ensures maximum absorption. (This is a secondary benefit, and not as significant as maximizing energy density.)

The angle at which sunlight energy strikes a surface of an array — or any object, for that matter — is known as the *angle of incidence*. (Technically, the angle of incidence is the angle between incoming solar radiation and a line perpendicular to the surface of the module.) Reducing the angle of incidence increases the irradiance and reduces the amount of sunlight that reflects off the array, increasing output. Ideally, the angle of incidence (angle of the incoming solar radiation) should be 0°.

Aligning the array at all times with the Sun to reduce the angle of incidence is possible through the use of trackers, which are actively or passively controlled solar mounting systems. As just noted, single-axis pole-mounted trackers automatically adjust for changes in the azimuth angle. That is, they track the Sun from east to west as it moves across the sky. They do not adjust the tilt angle as the altitude angle of the Sun changes during the day. Dual-axis trackers adjust the altitude *and* azimuth angles. That is, they track the Sun from sunrise to sunset, but also adjust the tilt angle of the array to accommodate changes in the altitude angle.

Although tracking helps improve the output of an array, most PV modules are mounted on fixed racks — set at a specific angle all year round for convenience. Fixed racks can be attached to poles or mounted on the ground or on the roofs of buildings. When installers mount a fixed rack, they adjust the tilt angle of the array to achieve optimum performance.

As you'll see in Chapter 8, most installers mount arrays at a tilt angle that corresponds to the latitude of the site. In Denver, Colorado, at 40° north latitude, installers typically mount arrays at a 40° tilt angle. In northern Florida, at 30° north latitude, they set the tilt angle of arrays at 30°.

Orienting a solar array as close to true south as possible with a tilt angle that matches the local latitude generally provides the best year-round performance for a fixed array. However, to optimize performance in the summer, the tilt angle should be reduced by about 15°, and to increase output in the winter, the tilt angle should be increased by about 15°.

Conclusion

Solar energy is an enormous resource that could heat and cool our homes and power lights and household electronic devices. It could even provide power to electric cars. To make the most of it, though, a solar array has to be oriented properly to take into account the Sun's daily and seasonally changing position in the sky. In the next chapter, I will turn my attention to the technologies that capture solar energy, solar cells.

CHAPTER 3

UNDERSTANDING
SOLAR ELECTRICITY

Although widespread interest in solar electricity is fairly recent, the technology owes its origin to researchers in the 1800s who found that light could move electrons in solid materials. This truly fascinating discovery led to the development of several types of solar cells that are in use today. In this chapter, we'll take a peek inside solar cells to see how they work. We examine the types of modules on the market today and introduce you to some important terms and concepts. For a history of solar cell technology and a discussion of new developments, you may want to check out my book, *Power from the Sun*.

What is a PV Cell?

Photovoltaic cells are solid-state electronic devices like transistors, diodes, and other components of modern electronic equipment. These devices are referred to as solid-state because electrons flow through solid material within them. Most solar cells in use today are made from one of the most abundant materials on the planet, silicon, which is extracted from quartz and sand.

Like all atoms, silicon atoms contain electrons that orbit around a central nucleus that contains protons and neutrons. In silicon, some of the electrons can be jolted loose from their orbit around the nuclei of the silicon atoms when struck by sunlight. These loose electrons can be made to flow together, creating an electrical current.

Because numerous solar cells are wired in series in a PV module, numerous electrons can be gathered up and conducted away from the array to power household loads.

Most solar cells in use today are thin wafers of silicon about $1/100^{th}$ of an inch thick (they range from 180 microns (μm) to 350 μm in thickness). As shown in Figure 3.1, most solar cells consist of two layers — a very thin upper layer and a much thicker lower layer. The upper layer is made of silicon and phosphorus atoms; the bottom layer consists of silicon and boron atoms.

When sunlight strikes the silicon atoms in solar cells, it jars electrons out of the atoms in both layers. These electrons flow

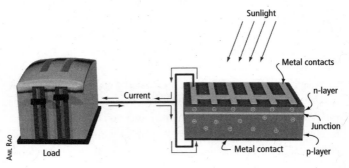

Fig. 3.1: *Cross Section through a Solar Cell. Solar cells like the one shown here consist of two layers of photosensitive silicon, a thin top layer, the n-layer, and a thicker bottom layer, the p-layer. Sunlight causes electrons to flow from the cell through metallic contacts on the surface of most solar cells, creating DC electricity. Solar-energized electrons then flow to loads where the solar energy they carry is used to power the loads. De-energized electrons then flow back to the solar cell.*

Understanding Silicon

Silicon PV cells are made from the element silicon. Silicon is refined from quartz, which is made of silicon dioxide, and from the type of sand that contains quartz particles.

preferentially toward the surface (for reasons beyond the scope of this book). These electrons flow into the metal contacts located on the front of solar cells. As noted in Chapter 1, numerous solar cells are wired in series in a solar module. Because of this, electrons extracted from one cell flow to the next cell, and then to the next cell, etc., until they reach the negative terminal of the module. Electrons flow from the array through wires connected to the house to power a load (any device that consumes electricity). After delivering the energy they gained from sunlight to the load, the de-energized electrons return through a different wire to the array. The electrons then flow back into the solar cells, filling the empty spots left in the atoms created by their ejection. This permits the flow of electrons to continue ad infinitum. For a more detailed description, you may want to check out *Power from the Sun*.

Types of PV Cells

Solar cells can be made from a variety of semiconductor materials. By far the most common is silicon. As noted earlier, silicon is produced from silicon dioxide, which is derived from two sources: quartzite and silica sand. Quartzite is a rock made entirely of the mineral quartz, which in turn consists of nearly pure silica (silicon dioxide). Silica sand is relatively pure sand containing a high percentage of silica. Geologically, silica sand is derived from quartz. Silicon dominates the semiconductor market even though there are other materials that more efficiently convert sunlight to electricity because silicon semiconductors produce the most electricity at the lowest cost.

Three forms of silicon are used to make solar modules: monocrystalline, polycrystalline, and amorphous.

Monocrystalline PVs

Monocrystalline cells — a.k.a. single crystal cells — were the first commercially manufactured solar cells. They are made from wafers sliced from a single large, cylindrical manufactured crystal of silicon (Figure 3.2). Single crystal ingots are made by melting highly purified

SOLARWORLD

Fig. 3.2: *Monocrystalline Silicon Ingot. This ingot is a huge crystal of silicon that is sliced to make monocrystalline PV cells.*

chunks of polysilicon and a trace amount of boron. Once melted, a seed crystal is dipped in the molten mass of silicon and boron. The seed crystal is rotated and slowly withdrawn. As it's withdrawn, silicon atoms from the melt attach to the seed crystal, exactly duplicating its crystal structure. Over time, the crystal grows larger and larger. Eventually, the ingot may grow to a length of 40 inches and a diameter of 8 inches.

Once extracted from the melt, the ingot is cooled. The rounded edges are trimmed and the square or rectangular ingot is then sliced with a diamond wire saw to produce ultrathin wafers used to make solar cells. Waste from this process is remelted and reused.

Monocrystalline cells boast the highest efficiency of all conventional PV cells — around 15% — although efficiencies vary from one manufacturer to the next, ranging from 14 to 17%.

Polycrystalline PV Cells

Polycrystalline solar cells are made from silicon with a trace of boron just like monocrystalline cells. To make a polycrystalline cell, however, the molten material is poured into a square or rectangular

Fig. 3.3: *Polycrystalline PV Cell. This close-up photo of a polycrystalline solar cell shows the individual crystals that form when molten silicon is cooled in the mold.*

mold. It is then allowed to cool very slowly. As the ingot cools, many smaller crystals form internally (Figure 3.3). Once cooled, the cast ingot is removed from the mold, then sliced using a diamond wire saw, creating wafers used to fabricate solar cells.

Polycrystalline solar cells are only about 12% efficient. (Their efficiency varies from one manufacturer to the next, ranging from 11.5 to 14%). Although polycrystalline cells are less efficient, they require less energy to produce. Because of this, they're a bit cheaper to manufacture.

Ribbon Polycrystalline PV

While most polycrystalline cells are sliced from ingots, wafers can also be made in long continuous ribbons. As shown in Figure 3.4b, a seed crystal is attached to two heat-resistant wires called filaments. The filaments are immersed in a molten mass of silicon and then slowly drawn from the melted silicon. The ribbon grows linearly as the molten silicon that spans the wires solidifies. Once the ribbon is completely formed, it is sliced into rectangular wafers.

Although the efficiency of ribbon silicon wafers is slightly lower than other PV cells — it falls in the 11 to 13% range — the technology offers several advantages. One of them is that it requires considerably less energy than monocrystalline and polycrystalline production. The ribbon technology also eliminates a lot of time-consuming, wasteful, and costly ingot slicing.

a

Fig. 3.4:

Ribbon Technology.

(a) Photo of solar cells in an Evergreen solar module.

(b) This drawing shows the ribbon making technology employed by Evergreen Solar.

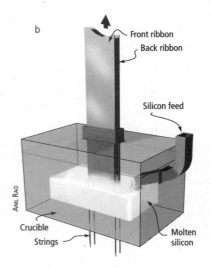

b

Front ribbon

Back ribbon

Silicon feed

Silicon
molten

Crucible

Strings

All three of the processes described so far result in the formation of the p-layer, the thicker, bottom layer of PV cells. How is the much thinner, phosphorus-containing, n-layer added?

After chemical preparation, the wafers are placed in a diffusion furnace and heated. Phosphorous gas is introduced into the furnace. In this high-temperature environment, phosphorus atoms penetrate the exposed top surface and sides of the wafer. (This results in the formation of a very thin, but complete n-type layer.

Metal contacts are then screen-printed onto the face of the wafer using a silver paste. Contacts are applied in a grid pattern consisting of two or more wide main strips and numerous ultra-fine (hair-thin) strips that attach perpendicularly to the main strips (Figure 3.5). This grid collects the electrons released from atoms inside the solar cell. As noted earlier, another conducting layer, usually made from aluminum, is applied to the back of each cell.

Because silicon is highly reflective, an anti-reflective coating is applied to the surface of the cells. It helps reduce the reflection of

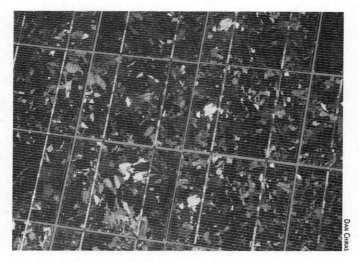

Fig. 3.5: *PV Cell Grid. These fine silver contacts described in the text draw electricity off the surface of the module.*

Energy Payback for PVs

You may have heard people say that it takes more energy to make a PV system than you get out of it over its lifetime. Fortunately, that's not even close to being accurate.

While it takes energy to make solar cells, modules, and the remaining components of a PV system, the energy payback is actually amazingly short — only one to two years, according to a study released in 2006 by CrystalClear, a research and development project on advanced industrial crystalline silicon PV technology funded by a consortium of European PV manufacturers (www.ipcrystalclear.info/default.aspx). As Justine Sanchez notes in her 2008 article in *Home Power*, "PV Energy Payback," "Given that a PV system will continue to produce electricity for 30 years or more, a PV system's lifetime production will far exceed the energy it took to produce it."

light off a module, resulting in greater absorption of the Sun's energy and greater output.

Thin-Film Technology

In an effort to produce solar modules at a lower cost — which means using less energy and less material — several manufacturers have turned to a new technology, known as *thin-film*. Unlike manufacturers of previous technologies, thin-film producers manufacture entire modules, rather than individual cells that must be wired into series to create modules. Skipping the step of assembling cells into modules saves energy, time, and money.

In silicon-based thin-film solar designs, known as *amorphous silicon*, silicon is deposited directly onto a metal backing (aluminum), glass, or even plastic by a technique known as *chemical vapor deposition*. This creates a thin film of photo-reactive material. Once it has solidified, a laser is used to delineate cells and create connections between the newly formed cells.

Thin-film PVs are currently made from amorphous silicon and three other blends of semiconductor materials. One of the advantages of amorphous silicon, notes Erika Weliczko, in an article in *Home Power* (Issue 127), is that it can be manufactured in long, continuous rolls, or incorporated onto flexible substrates such as laminates, shingles, and roofing.

Efforts are being made to increase the efficiency of this promising technology. By carefully selecting the semiconductors, manufacturers can create solar modules that absorb much more of the solar spectrum. The result is a more efficient PV cell — one that converts more of the Sun's energy into electricity.

Thin-film technology offers several advantages over single and polycrystalline solar cells. One of the most important is that their production uses considerably less energy by eliminating the costly and energy-intensive ingot production and wafer slicing required in manufacturing mono- and polycrystalline PV cells. Another advantage is that it uses less material. Although silicon is abundant, it is not cheap. In recent years, supplies of this semiconductor have been limited because of competing demand from the computer industry, which uses silicon to make computer chips. Another advantage of thin-film PV is that it is less sensitive to high temperatures. At 100°F, a crystalline module will experience a 6% loss in production while a thin-film amorphous silicon array will experience a 2% loss, making thin-film a good choice for sunny climates, provided you have the roof space. Thin-film is also a bit more shade tolerant than crystalline PV. One of the chief disadvantages of thin-film PVs is their low efficiency. Most thin-film products have efficiencies in the 6 to 8% range. Lower efficiency means larger arrays. In fact, you would need approximately 180 square feet of roof space for a 1 kW PV array made from thin-film vs. 90 square feet for a comparable array made from crystalline PV modules.

Rating PV Modules and Sizing PV Systems

PV manufacturers rate modules by various parameters, including rated power, power per square foot, and efficiency. (Additional parameters are discussed in *Power from the Sun* for those wishing to know more.) These parameters are used to compare PV modules and to design PV systems to ensure that components can handle the voltage and current produced by the array under all conditions.

Rated power is the wattage a module produces under standard test conditions (STC). This measurement is used to compare one module to another. It is also useful when a professional or do-it-yourselfer is sizing a PV array. As noted in Chapter 1, the standard test conditions used to determine rated power rarely reflect typical operating conditions. A more practical measure is the rated power derived under more typical field conditions, known as PTC. (For more on this see the section "Rated Power and Capacity" in Chapter 1 on page 3).

Some manufacturers also list *rated power (watts) per square foot* of module. As the name implies, it is the power output per square foot of module area. This number is useful to those who have a limited amount of space to mount a PV array and want to generate as much electricity as possible in that limited space.

Module efficiency is also a handy number. It's the ratio of output power (watts) from a module to input power (watts) from the Sun. A 15% module efficiency means that 15% of the incident (incoming) solar radiation is converted into electricity. Like rated power per square foot, it's handy for those who have limited space.

Advancements in PVs — What's on the Horizon?

Although PVs have come a long way since the first work in the early 1800s, researchers in the private and public sectors continue to look for ways to improve the efficiency and reduce the cost of solar electric technologies to make solar electricity more affordable to customers and profitable for businesses. This effort has led to some rather promising developments in recent years. Some of these technologies are already available; others are not quite ready for prime time, but could be manufactured commercially in the near future.

One of the newest and most popular PV technologies to make its way to the market is *building-integrated PV* (BIPV) (Figure 3.6).

BIPV incorporates solar electric generating capacity into the components of a building envelope, for example, the roof, windows, skylights, and exterior walls.

BIPVs are being used primarily in new buildings, although they can be incorporated into existing homes and offices. They offer several advantages over more conventional PV technologies. One of the most important advantages is that they perform multiple functions, which reduces resource depletion and construction costs. For instance, solar shade structures, like the one shown in Figure 3.6, provide shade for windows, lower cooling costs, and eliminate the need for conventional awnings. They do all this while generating electricity.

Another advantage of BIPVs is that they tend to blend in better than conventional PV modules, which many people find appealing. Because of these advantages, BIPV is now one of the fastest-growing parts of the PV industry.

Another new PV technology is the *dye-sensitized solar cell* (DSSC).

DSSCs, unlike previous cells described in this chapter, are not solid-state devices. They are photoelectrochemical cells. Although

Fig. 3.6: *Solar deck shade. Solar shade structures, like the one shown here, provides shade for windows and walls, eliminating the need for conventional awnings. Like other forms of building integrated PVs, they perform more than one function. Awnings shade windows and walls and thus cool buildings in the summer and produce electricity year round.*

LIGHTHOUSE SOLAR IN BOULDER, CO

they rely on the photoelectric effect to generate electrons (in the dye) to create current, they also rely on ions in the electrolyte to transfer electrons to the dye molecules.

This technology uses low-cost materials and can be used to manufacture flexible sheets, as in thin-film technology. Moreover, cells can be assembled using much less energy and much simpler and less expensive equipment than is required to manufacture conventional monocrystalline and polycrystalline PVs. As a result, they should be cheaper to manufacture. Lower production costs could enable these cells to compete with electricity generated by coal and other fossil fuels within the next decade or so. DSSCs are starting to make their way to the markets as battery chargers mounted on sports bags and back packs.

Another technology that holds promise for the future is the organic or polymer solar cells. Organic solar cells consist of thin plastic films (typically 100 mm) containing chemicals that release electrons when struck by light. Organic cells are inefficient — so far, the highest efficiencies are around 6.5% — but they could potentially be produced very cheaply, according to some experts.

Yet another promising technology is the bifacial solar cell, also known as a hybrid solar cell. According to Sanyo, manufacturer of a bifacial module, bifacial solar cells consist of a monocrystalline silicon wafer sandwiched between two ultra-thin amorphous silicon layers. Because of this, these modules can capture sunlight falling on both sides of their cells. (This feature is the reason they're called bifacial modules.)

In a bifacial module, the front side of the panel generates electricity from direct and diffuse solar radiation (Figure 3.7). The backside generates electricity from diffuse light from the sky as well as reflected light from surrounding surfaces. Under the right conditions, bifacial modules can produce more power than conventional single-sided modules.

The efficiency of solar modules can be increased by using lenses and mirrors to concentrate sunlight on PV cells (Figure 3.8). Concentrating sunlight greatly increases the solar input to the cells.

3.7 a

3.7 b

Fig. 3.7 a and b: *Bifacial Modules.*
*(a) Bifacial modules harvest solar
energy off both sides of the panel,
although the front side shown in this
photograph is by far the most productive source of electricity.*
*(b) Sunlight reflecting off the light-
colored roofing illuminates the array,
boosting its output.*

Fig. 3.8: *Concentrating Solar Collectors. Solar collectors like these large
commercial arrays use lenses to concentrate sunlight on PV cells, dramatically
boosting their output.*

The more sunlight that can be focused on a PV, the more electricity it generates. Note that most concentrating collectors are mounted on trackers — devices that follow the Sun across the sky — to operate properly.

Conclusion: Should You Wait for the Latest, Greatest New Technology?

Knowing that new PV technologies are in the offing, many people ask if they should wait a bit before they invest in a solar electric system. Does it make sense to delay your installation until the new, more efficient PV technologies hit the market?

This is a fair question, but the answer is no.

Newer, more efficient PVs are certainly on their way; however, one of the key considerations when installing a PV system is not the efficiency of the PV modules, but the cost of the modules based on the installed capacity. This is referred to as the *cost per watt of installed capacity.*

When comparing modules on this basis, you will find that new, more efficient technologies typically cost more per watt of installed capacity. A 2 kW system that utilizes the most efficient modules on the market, for instance, may cost 10 to 30% more to purchase than a 2 kW system that uses slightly less efficient modules. Yet both arrays produce the same amount of electricity. Why pay more to produce the same amount of electricity? If space is limited, efficiency matters. If meeting your electrical needs requires a 3 kW system and the only place with good solar access you have to mount the modules is on a small garage roof, you may need to install the more efficient — and expensive — modules.

While conversion efficiency may not matter to you, efficiency is driving the market, and over the long haul it will result in more efficient and (hopefully) less expensive modules.

CHAPTER 4

IS SOLAR ELECTRICITY
RIGHT FOR YOU?

Before you invest your hard-earned money in a solar electric system, it is important to determine whether a solar electric system makes sense for you. Will a PV system meet your needs? How much will it cost? Would you be better off using utility power? Or would some other type of renewable energy system make more sense? And even if it doesn't make perfect sense economically, what about the environmental benefits and personal gratification of doing the right thing? What about alternatives to buying your own PV system, in particular, leases and power purchase agreements? In this chapter, I'll help you answer these questions.

Assessing Electrical Demand

The cost of a solar electric system depends on many factors, among them the size of the system, the complexity of the system, the distance the installer must travel, the type of installation, and the difficulty of the installation.

Because solar resources and electrical consumption vary, most residential solar electric systems fall within the 1 to 6 kW range, the most common being 3 to 6 kW systems. In an all-electric home equipped with a wide assortment of electric appliances — such as central air conditioners, electric space heaters, electric water heaters, and electric stoves — average monthly electrical consumption typically falls within the 2,000 to 3,000 kilowatt-hours/month range. Even in very sunny climates, very large PV systems would be required to meet electrical demand — systems on the order of 10 kW.

In homes in which natural gas or propane is used to cook food, heat water, and provide space heat, electrical consumption may be as low as 400 or 500 kWh per month. A much smaller system would be needed in such instances. In a relatively sunny climate, a 3 to 4 kW system might suffice.

Although many factors affect the cost of a PV system, the size of the system depends on electrical consumption and solar resources. How you go about calculating consumption depends on whether it is an existing structure or one that's about to be built. Let's begin with existing structures.

Assessing Electrical Demand in Existing Structures

Assessing the electrical consumption of an existing home or business is fairly easy. Most people obtain this information from their monthly electric bills, going back two to three years, if possible. (In some areas, every utility bill includes a summary of year-to-date electrical consumption.) If you don't save your electric bills, a call to the local power company will usually yield the information you need. Some customers can access the data online through their utility company's website. All you need is your customer number.

If you purchased a home that's been around for a while, you can obtain energy data from the previous owner. Remember, however, a house does not consume electricity, its occupants do, and we all use energy differently. If a previous homeowner and his or her family used energy wastefully, their energy consumption data may be of little value to you if you and your family are energy misers.

To determine total annual electrical consumption from utility bills, don't look at the cost in dollars and cents; look instead for the kilowatt-hours of electricity consumed each month. Calculate the total for each year and then calculate a yearly average. If you're considering installing a grid-tied solar electric system — a PV system that feeds surplus electricity back to the grid — annual electrical demand is often all you need. You can size your system based on this information. If you are thinking about severing your ties with the utility — that is, taking your house off the grid (a step I don't typically recommend) — you will need to calculate monthly averages as well — that is, how much electricity is used, on average, during each month of the year. To do this, add up household electrical consumption for each month, and then divide by the number of years' worth of data you have. For example, if your records go back four years, add the electrical consumption for all four Januarys, and then divide by four. Do the same for February and each of the remaining months. This will permit you to determine when electrical demand is the greatest. Off-grid systems are sized to meet demands during the times of highest consumption.

After you have calculated energy consumption, take a few moments to look for trends in energy use. Is energy consumption on the rise or is it staying constant or declining? If energy use is rising, you should dump earlier years' data. It doesn't reflect your current consumption. If you don't, you'll end up undersizing your system.

If, on the other hand, electrical energy consumption has declined because you've become more energy efficient, earlier data will artificially inflate electrical demand. You'll need less electricity than the averages indicate and a smaller system.

Assessing Electrical Demand in New Buildings

Determining monthly electrical consumption is fairly straightforward in existing homes and businesses. Determining electrical demand in a new home — either one that's just been built or one that is about to be built — is much more difficult.

One method used to estimate electrical consumption is to base it on the electrical consumption of your existing home or business. If, for instance, you are building a brand new home that's the same size as your current home and the new home will have the same amenities and the same number of occupants, electrical consumption could be similar to your existing home.

If, however, you are building a more energy-efficient home and are installing much more energy-efficient lighting and appliances and incorporating passive solar heating and cooling (all of which I highly recommend), electrical consumption could easily be 50%, perhaps 75% lower than in your current home. If that's the

Table 4.1 Electrical Consumption Chart					
Individual Loads	Qty X	Volts X	Amps =	Watts AC	DC

AC Total Connected Watts: _____

DC Total Connected Watts: _____

case, you can adjust electrical demand to reflect the efficiency upgrades.

Another way to estimate electrical consumption is to perform a load analysis. A load analysis is an estimate of electric consumption based on the number of electronic devices in a home, their

	X Use Hrs/day	X Use days/wk	÷ 7 days	= Watts Hours AC	DC
			7		
			7		
			7		
			7		
			7		
			7		
			7		
			7		
			7		
			7		
			7		
			7		
			7		
			7		

AC Average Daily Load: _____

DC Average Daily Load: _____

Plug-in Watt-Hour Meters

The Kill A Watt meter or the Watts Up? meter make it easy to measure power (watts) and energy (watt-hours) used by small devices. Just plug the meter into an outlet and plug the appliance into the meter. These meters read real power (watts), apparent power (volt amps), power factor, volts, amps, and elapsed time (used for calculating average power use over time).

Power meters are useful for estimating load, and also for gaining an appreciation for the energy demands of household appliances. Knowing which appliances and devices are the energy hogs in a home may empower occupants to reduce energy consumption. Surveying the energy use of household appliances could be a great educational project for children.

Both meters are available in a range of models, some with advanced features. The basic model Kill A Watt meter list price is about $35; the basic model Watts Up? meter list price is about $96. Both are available at discount. As an alternative to purchasing a meter, some public libraries and utilities have plug-in watt-hour meters available for loan.

average daily use, and energy consumption. To perform a load analysis, a homeowner begins by listing all the appliances, lights, and electronic devices in his or her new home or office. To keep track of all loads, I recommend using a worksheet like the one shown in Table 4-1. It can be found online.

Once you've prepared a complete list of all the devices that consume electricity, your next assignment is to determine how much electricity each one uses. This can be done by referring to the name plate on the back side of each electronic device. It lists wattage or sometimes just the amps. To determine watts, multiply amps by 120 volts. You can also determine consumption by directly measuring each device with a watt meter like the ones shown in Figure 4.1. (The accompanying sidebar describes these devices.)

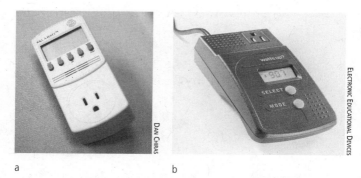

a   b

Fig. 4.1: *Kill A Watt and Watts Up? Meters. (a) The Kill A Watt meter and (b) Watts Up? meter shown here can be used to measure wattage of household appliances and electronic devices. Of the two, the Watts Up? is the more sensitive and allows for measurement of tiny phantom loads as well.*

After you have determined the wattage of each electrical device and light, you must estimate the number of hours each one is used on an average day and how many days each device is used during a typical week. From this information, you calculate the weekly energy consumption of all devices in your home or business. You then divide this number by seven to determine the average daily consumption of your home or business in watt-hours. You will use this number to determine the size of your PV system.

Load analysis may seem simple at first glance, but it is fraught with problems. (For more details on this process, you may want to read more about it in *Power from the Sun*.)

One problem with load analysis is that many electronic devices draw power when they're off. Such devices are known as a *phantom loads*. They include television sets, VCRs, satellite receivers, cell phone and laptop computer chargers, and a host of other common household devices. A few, like satellite receivers, draw nearly as much power when they're off as when they're on (Figure 4.2). Phantom loads typically account for 5 to 10% of the monthly electrical consumption in US homes. If they're not factored into the load analysis, estimates can be off.

DAN CHIRAS

Fig. 4.2: *Phantom Load. Unbeknownst to most of us, our homes are filled with phantom loads, devices that draw power when not in use, like this television set and satellite receiver. Of this duo, the satellite receiver is by far the worst offender. It uses 15 watts when on and 14 watts when turned off. Phantom loads account for about 5 to 10% of a home or business' annual electrical demand.*

Once you've calculated daily electrical energy use, the next step is to find ways to reduce demand.

Reducing energy demand, reduces the size of the solar electric system you'll need to install. The smaller the system, the less you'll spend to purchase and install your system. How much can you save?

Richard Perez, founder of *Home Power* magazine, asserts that every dollar invested in energy efficiency saves $3 to $5 in the cost of a solar electric system. For instance, if you invest $2,000 in measures that trim the electrical energy use in your home or office, for instance, by sealing leaks, adding insulation, and installing more efficient lighting, you could save $6,000 to $10,000 on the cost of a solar electric system.

Because efficiency is both economically and environmentally superior to increasing the capacity of a renewable energy system, most reputable solar installers recommend energy conservation and energy efficiency measures first. If you agree to pursue these measures, they downsize your system and save you thousands of dollars. Solar installers may make specific recommendations. Solar site assessors do the same. In fact, they're trained to offer objective, third-party advice on efficiency and solar system requirements. (To locate a solar PV site assessor in your area, visit the Midwest Renewable Energy Association's website.)

Another option is to hire a home energy auditor. A qualified home energy auditor will perform a more thorough energy analysis of your home and make recommendations on ways you can reduce your demand. They will also prioritize energy-saving measures. Or, you can perform your own home energy audit. For guidance, you may want to check out one of my newest books, *Green Home Improvement*. This book will teach you how to perform an audit and also contains several dozen home energy efficiency projects you can perform.

Sizing a Solar Electric System

Once you have estimated your electrical demand and implemented a strategy to use energy more efficiently, it is time to determine the size of the system you'll need to meet your needs. This is a step usually performed by professional solar electric installers. The size of the system varies, of course, depending on electrical demand and your goals. Although I'll elaborate on system options in the next chapter, for now, it's important to note that PV systems fall into three broad categories: (1) grid-connected, (2) grid-connected with battery backup, and (3) off-grid.

A grid-connected or utility-connected solar electric system can be sized to meet some or all of the electrical needs of a home or business. Excess electricity, if any, is fed back onto the grid, running the electric meter backward. These systems are by far the most popular. A grid-connected system with battery backup is similar to a grid-connected system, but has a battery bank to store electricity. These systems will continue to supply critical loads if the grid goes down.

As their name implies, off-grid PV systems are not connected to the utility grid. They're completely autonomous electric-generating systems. Surpluses are stored in batteries for use at night or on cloudy days.

Sizing a Grid-Connected System

Sizing a grid-connected system is the easiest of all. To meet 100% of your needs, simply divide your average daily electrical demand (in kilowatt-hours) by the average peak sun hours per day for your area. As noted in Chapter 2, peak sun hours can be determined from solar maps and from various websites and tables.

As an example, let's suppose that you and your family consume 6,000 kWh of electricity per year (or 500 kWh per month). This is about 17 kWh per day. Let's suppose you live in Lexington, Kentucky where the average peak sun hours per day is 4.5. Dividing 17 by 4.5 gives you the array size (capacity): 3.8 kilowatts. But don't run out and order a system based on that number.

This calculation yields system size if the system were 100% efficient and unshaded throughout the year. Unfortunately, no PV system is 100% efficient. As a result, most solar installers derate grid-connected PV systems by 22-25% to account for losses. This loss is due to voltage drop as electricity flows through wires; resistance at fuses, breakers, and connections; dust on the array; and inefficiencies of various components such as the inverter. So, if your PV array is not shaded by trees or nearby buildings, you'd need a 22-25% larger system to provide 17 kWh of electricity per day. For the example above, you would calculate the proper array size by dividing 3.8 kW by 0.78 - 0.75 to account for inefficiencies. The result is 4.87 kW *provided the solar array is not shaded.*

Shading — even a small amount — dramatically lowers the output of a PV system. To determine the amount of shading on a PV system, professional solar site assessors and solar installers often use a sun path analysis tool like the Solar Pathfinder, which is shown in Figure 4.3. The Solar Pathfinder and similar devices determine the percentage of solar radiation blocked by permanent

Fig. 4.3:
*Solar Pathfinder.
This device allows solar
site assessors and
installers to assess
shading at potential
sites for PV systems,
helping you select the
best possible site to
install an array.*

Fig. 4.4:
*The dome of the Solar
Pathfinder shows all
sources of shade
throughout the year, no
matter when you use the
device. This allows solar
site assessors to deter-
mine how much shading
occurs and when, so an
array's output can be
accurately estimated.*

local features in the landscape such as trees, hills, and buildings
(Figure 4.4). (How they're used is discussed in *Power from the Sun.*)
This data is then used to calculate the final size. You'll need to
increase the array size to take into account any shading.

Solar installers can also size a PV system based on budget — how much money a person has to spend — because not everyone can afford a PV system that meets 100% of their needs.

For optimum performance, PV systems generally require a site where an array is unshaded at least three hours on either side of solar noon — that is, from 9 AM to 3 PM solar time. This portion of the daily solar window accounts for 75 to 80% of the Sun's daily radiation, depending on the season.

Sizing an Off-Grid System

As a rule, off-grid systems are sized according to the month with the highest demand, which is often the month with the least sunshine. In temperate climates — for example, in states like Wisconsin, Kansas, New York, and Oregon — the array is typically sized using insolation data from November, December, or January. (These are the months with the highest electrical demand and are also the shortest and typically the cloudiest months of the year in cold climates.) First, determine the month with the highest demand. Then divide the average daily consumption by the average peak sun hours during that month to determine the size of the array, then adjust for efficiency and shading, as described previously.

The size of the battery bank must also be carefully calculated. Battery banks are sized to provide sufficient electricity to meet a family's or business' needs during cloudy periods. Most battery banks are based on a three to five day reserve, so they can provide enough electricity for three to five days of cloudy weather. Battery banks are also sized to prevent deep discharging during this period — that is, so the batteries do not discharge more than 50% of their capacity. Deep discharging can seriously damage batteries.

Sizing a Grid-Connected System with Battery Backup

PV arrays in grid-connected systems with battery backup are sized much like grid-connected systems, discussed earlier. As a rule, the size of the array is based on expectations, that is, the percentage of electrical demand you'd like your system to meet, taking into account peak sun hours, efficiency, and shading.

Battery banks are sized according to the electrical requirements of *critical loads* — that is, the electrical load that would be needed during a power outage. Critical loads are usually restricted to pumps or fans of heating or cooling systems, well pumps, sump pumps, refrigerators, and a few lights. Indispensible medical equipment may need to be included as well. Because these loads require less energy than the entire home, batteries are typically much smaller in grid-connected systems with battery backup than off-grid systems.

Does a Solar Electric System Make Economic Sense?

At least three options are available to analyze the economic costs and benefits of a solar electric system: (1) a comparison of the cost of electricity from the solar electric system with conventional power or some other renewable energy technology, (2) an estimate of return on investment, and (3) a more sophisticated economic analysis tool known as total cost of ownership.

Cost of Electricity Comparison

One of the simplest ways of analyzing the economic performance of a solar system is to compare the cost of electricity produced by a PV system to the cost of electricity from a conventional source such as the local utility. This is a five-step process, two of which I've already discussed.

The first step is to determine the average monthly electrical consumption of your home or business, preferably after incorporating conservation and efficiency measures. Second, calculate the size of the system you'll need to install to meet your needs. Third, calculate the cost of the system. (A solar provider can help you with this.) Fourth, after determining the cost of the system, calculate the output of the system over a 30-year period, the expected life of the system. Fifth, now estimate the cost per kilowatt-hour by dividing the cost of a PV system by the total output.

Suppose you live in Colorado and are interested in installing a grid-connected solar electric system that will meet 100% of your electric needs. Your super-efficient home requires, on average, 500 kWh of electricity per month. That's 16.4 kWh per day. Peak sun hours is 6. To size the system, divide the electrical demand (16.4 kWh per day) by the peak sun hours. The result is a 2.7 kW system. Adjusting for 78% efficiency, the system should be 3.46 or 3.5 kW. For the sake of simplicity, let's assume that the system is not shaded at all during the year.

Your local solar installer says she can install the system for $7 a watt, or $24,500. You'll receive a rebate from the utility of $3.50 per watt of installed capacity or $12,250. The system cost is now $12,250. You'll also receive a 30% tax credit from the federal government on the cost of the system. The federal tax credit is based on the initial cost of the system ($24,500) minus the utility rebate ($12,250 in this example). Thirty percent of this amount equals $3,675. Total system cost after subtracting these incentives is $8,575.

According to your calculations or the calculations provided by the solar installer, this system will produce about 16.4 kWh of electricity per day or 6,000 kWh per year. If the system lasts for 30 years, it will produce 180,000 kWh.

To calculate the cost per kilowatt-hour, divide the system cost ($8,575) by the output (180,000 kWh). In this case, your electricity will cost slightly less than 4.76 cents per kWh. Considering that the going rate in Colorado is currently about 9.5 cents per kWh, including all fees and taxes, the PV system represents a pretty good investment.

Return on Investment

Another relatively simple method used to determine the cost effectiveness of a PV system is simple return on investment (ROI). Simple return on investment is, as its name implies, the savings generated by installing a PV system expressed as a percentage of the investment.

Simple ROI is calculated by dividing the annual dollar value of the energy generated by a PV or wind system by the cost of the system. A solar electric system that produces 6,000 kWh of electricity per year in an area where electricity sells at 9.5 cents per kilowatt-hour generates $570 worth of electricity each year. If the system costs $8,575, after rebates, the simple return on investment is $570 divided by $10,710 x 100 which equals 6.6%. If the utility charges 15 cents per kWh, the 6,000 kWh of electricity would be worth $900 and the simple ROI would be 10.5%. Given the state of the economy, both of these represent decent rates of return. (If only our retirement funds performed half as well these days!) Even in good economic times, these are respectable ROIs.

Weaknesses of Economic Analysis Tools

Comparing the cost of electricity and return on investment are both simple tools. Both fail to take into account a number of economic factors. For example, both techniques fail to account for interest payments on loans that may be required to purchase a PV system. Interest payments will add to the cost of electricity produced by the system. For those who self-finance, for example, by taking money out of savings, both tools fail to take into account opportunity costs — lost income from interest-bearing accounts raided to pay for the system.

Both methods fail to take into account the rising cost of electricity. Nationwide, electric rates have increased on average about 4.4% per year over the past 35 years. In recent years, the rate of increase has been double that amount in some areas.

Although these analytic tools fail to account for key economic factors that would decrease the value of a PV system, the rising cost of electricity from conventional sources will in all likelihood offset the opportunity cost or the cost of financing a system.

That said, these techniques do not take into account system maintenance, insurance, or property taxes, if any. All of these factors add to the cost of a system over the long term.

Despite these shortcomings, the cost comparison and simple return on investment are convenient tools for evaluating the economic performance of renewable energy systems. They're infinitely better than the old standby, *payback* (also known as "simple payback").

Why?

Payback is a term that gained popularity in the 1970s. It was used to determine whether energy conservation measures and renewable energy systems made economic sense. Payback is the number of years it takes a renewable energy system or energy efficiency measure to pay back its cost through the savings it generates.

Payback is calculated by dividing the cost of a system by the anticipated annual savings. If the $8,575 PV system I've been looking at produces 6,000 kilowatt-hours per year and grid power costs you 9.5 cents per kWh, the annual savings of $570 yields a payback of 18.8 years ($8,575 divided by $570 = 15 years). In other words, this system will take 15 years to pay for itself. From that point on, the system produces electricity free of charge.

While the payback of 15 years on this system seems long, don't forget that the return on investment on this system, calculated earlier, was 6.6%, which is a very respectable rate of return on your investment — or any investment these days.

While simple payback is fairly easy to understand, it has some very serious drawbacks. The most important is that payback is a foreign concept to most of us and, as a result, can be a bit misleading.

Besides being misleading, simple payback is a concept we rarely apply in our lives. Do avid anglers ever calculate the payback on their new bass boats? ($25,000 plus the cost of oil, gas, and transportation to and from favorite fishing spots divided by the total number of pounds of edible bass meat at $5 per pound over the lifetime of the boat.) Do couples ever calculate the payback on their new SUV or the new chandelier they installed in the dining room?

Simple payback and simple return on investment are closely related metrics. In fact, ROI is the reciprocal of payback. That is, ROI = 1/Payback. Thus, a PV system with a 10-year payback represents a 10% return on investment (ROI = 1/10). A PV system with a 20-year payback represents a 5% ROI.

Although payback and ROI are related, return on investment is a much more familiar concept. We receive interest on savings accounts and are paid a percentage on mutual funds and bonds — both of which are a return on our investment. Most of us were introduced to return on investment very early in life — when we opened our first interest-bearing account. Renewable energy systems also yield a return on our investment, so it is logical to use ROI to assess their economic performance.

Adjusting for Incentives

When calculating the cost of electricity from a solar electric system, be sure to remember to subtract financial incentives from state and local government or local utilities — as in the previous example. Financial incentives can be quite substantial. In Wisconsin, for example, more than 30 utilities participate in a statewide program called Focus on Energy through which they provide customers who install PV systems a rebate of up to 25% of their system cost with a maximum reward of $35,000. Other utilities and even several states, like New York, offer generous incentives as well. The best PV incentives are found in Colorado, New Jersey, Massachusetts, California, and Oregon.

The federal government also offers a generous financial incentive to those who install PV systems. Their incentive is a 30% tax credit to homeowners and businesses. However, the feds also allow businesses to depreciate a solar electric system on an accelerated schedule, which means they can deduct the costs faster than other business equipment, recouping their investment more quickly. This further reduces the cost of a PV system. The US Department of Agriculture offers a 25% grant to cover the cost of PV systems on farms and rural businesses. Their minimum grant is $2,500 (for a $10,000 system) and the maximum is $500,000. To learn more about state and federal incentives in your area, log on to the Database of State Incentives for Renewables and Efficiency at www.dsireusa.org. Click on the map of your state. To learn more about USDA grant program, log on to www.rurdev.usda.gov/.

Because financial incentives can reduce the cost of a PV system, most PV system installations are driven by incentives.

Discounting and Net Present Value: Total Cost of Ownership

For those who want a more sophisticated tool to determine whether an investment in solar energy makes sense, economists offer up *discounting* and *net present value*. This method, referred to as total cost of ownership, allows you to compare the cost of a PV system to the cost of the electricity it will displace. Unlike the previous economic tools, this one takes into account numerous economic parameters, including initial costs, financial incentives, maintenance costs, the rising cost of grid power, and another key element, the time value of money. The time value of money takes into account the fact that a dollar today is worth more than a dollar tomorrow and even more than a dollar a few years from now. Economists refer to this as the *discount factor*.

To make life easier, this economic analysis can be performed by using a spread sheet. This method is discussed in *Power from the Sun*.

Alternative Financing for PV Systems

Not everyone has access to the money required to purchase a PV system — even with incentives — or wants to incur that kind of debt. If you are one of these people, there are some alternative financing mechanisms that could still make your dreams of a PV system come true: *power purchase agreements* (PPAs) and leases.

In a power purchase agreement, a private company installs a solar electric system on a customer's home — mostly at their expense. (They do require a down payment to help offset the cost of the system and installation.) The company owns and "operates" the system, selling the electricity generated by the system to the homeowner at a low rate — usually a rate that increases much more slowly than utility rates for the duration of the lease, typically around 18 years. Homeowners benefit because they incur no upfront costs while enjoying lower electric bills and living a more environmentally friendly lifestyle. They also own a residence that will probably sell more quickly when the time comes to put it on the market.

Another option is a lease. Once again, the PV system is installed by a private company and the system is leased to them. Customers that lease PV systems typically end up paying slightly less for electricity. The lease also guarantees a fixed rate for the term of the agreement, providing a hedge against rising electric rates.

Lease programs are available in California, Arizona, Oregon, Colorado and Connecticut. Expect to see other companies enter the market in other states.

Lease programs and power purchase agreements are really quite similar. The main difference is that in lease programs there's typically no down payment. However, as author and market analyst Charles W. Thurston explains in an article in *Home Power* magazine (issue 128), "if you can afford to invest up front in part of the system cost (through a PPA), you'll pay less as time goes on, and your savings can be greater at the end of the contract. In that case, a PPA may be more beneficial."

Despite Thurston's analysis, representatives from both industries argue that the financial costs are not that different over the long haul. "The bottom line is that a solar lease or PPA makes it possible for any homeowner to stop talking about tomorrow and act now," says Thurston. If you'd like to power your home with solar electricity, but can't afford a system or don't want to borrow the money, consider a lease or a power purchase agreement.

Putting It All Together

In this chapter, you've seen that there are several ways to save money on a PV system. Efficiency measures lower the initial size and cost of a system, saving huge sums of money. Tax incentives and rebates also lower the cost. Some states exempt PV systems from sales taxes or property taxes, creating additional savings. I encourage those who are building superefficient passive solar/solar electric homes to view savings they'll accrue from efficiency measures and passive solar design as a kind of internal subsidy or rebate for their

PV systems. My own solar electric system cost about $17,000 and has generated about $4,000 worth of electricity in the first 12 years. The return on investment is pretty low. However, my passive solar home has saved me approximately $18,000 in heating bills during this same period. Savings on electricity from the PV and savings on heating bills resulting from passive solar heating have more than paid for my PV system.

Economics is where the rubber meets the road. Comparing solar electric systems against the "competition" and calculating the return on investment gives a potential buyer a much more realistic view of the feasibility of solar energy at a particular site. Remember, however, economics is not the only metric on which we base our decisions. Energy independence, environmental values, reliability, the cool factor, bragging rights, the fun value, and other factors all play prominently in our decisions to invest in renewable energy.

CHAPTER 5

SOLAR ELECTRIC SYSTEMS —
WHAT ARE YOUR OPTIONS?

P V systems fall into three categories: (1) grid-connected, (2) grid-connected with battery backup, and (3) off-grid. In this chapter, We'll examine each system and then explore hybrid renewable energy systems — those that couple PV electric systems with other renewable energy sources. The information in this chapter will help you decide which system suits your needs, lifestyle, and pocketbook.

Grid-connected PV Systems

Grid-connected PV systems are the most popular solar electric system on the market today. As shown in Figure 5.1, grid-connected systems are so named because they are connected directly to the electrical grid — the vast network of electric wires that spans the nation and crisscrosses your neighborhood. These systems are also sometimes referred to as *batteryless grid-connected* or *batteryless utility-tied* systems.

As shown in Figure 5.1, a grid-connected system consists of five main components: (1) a PV array, (2) an inverter, (3) the main service panel or breaker box, (4) safety disconnects, and (5) meters.

To understand how a batteryless grid-connected system works, let's begin with the PV array. The PV array produces DC electricity. It flows through wires to the *inverter*, which converts the DC electricity to AC electricity. (For a description of AC and DC electricity, see the sidebar "AC vs. DC Electricity.")

a

A Inverter
B Breaker box
   (main service panel)
C Utility meter
D Wire to utility line
E Circuits to household loads

b

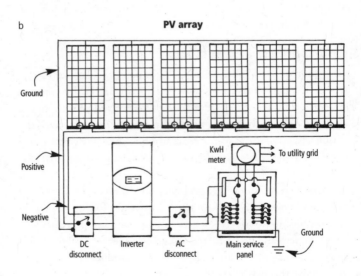

Fig. 5.1 a and b: *Schematic of Grid-connected PV System. (a) overview and (b) drawing of system wiring and system components.*

AC vs. DC Electricity

Electricity comes in two basic forms: *direct current* and *alternating current*. Direct current electricity consists of electrons that flow only in one direction through an electrical circuit. It's the kind of electricity produced by a flashlight battery or the batteries in portable devices like cell phones and laptop computers. It is also the kind of electricity produced by photovoltaic modules.

Most other sources of electricity, including wind turbines and conventional power plants, produce alternating current electricity. Like DC electricity, AC electricity consists of the flow of electrons through a circuit. However, in alternating current, the electrons flow back and forth. That is, they change (alternate) direction in very rapid succession, hence the name. Each change in the direction of flow (from left to right and back again) is called a cycle.

In North America, electric utilities produce electricity that cycles back and forth 60 times per second. It's referred to as 60-cycle-per-second — or 60 Hertz (Hz) — AC. In Europe and Asia, the utilities produce 50-cycle-per-second AC.

The inverter doesn't just convert the DC electricity to AC, it converts it to grid-compatible AC — that is, 60 cycles per second, 120-volt (or 240-volt) electricity. (See sidebar "Frequency and Voltage" for more on this.) Because the inverter produces electricity in sync with the grid, inverters in these systems are often referred to as *synchronous* inverters.

The 120-volt or 240-volt AC produced by the inverter flows to the main service panel, a.k.a. the breaker box. From there, it flows to active loads — that is, to electrical devices that are operating. If the PV system is producing more electricity than is needed to meet these demands — which is often the case on sunny days — the excess automatically flows onto the grid.

As shown in Figure 5.1, surplus electricity travels from the main service panel through the utility's electric meter, typically mounted

Frequency and Voltage

Alternating current is characterized by two parameters, frequency and voltage. Frequency refers to the number of times electrons change direction every second and is measured as cycles per second. (One cycle occurs when the electrons switch from flowing to the right then to the left then back to the right again.) The flow of electrons through an electrical wire is created by a force. Scientists refer to this force as *voltage*. The unit of measurement for voltage is *volts*.

Voltage is a more difficult electrical term to understand. You can think of it as electrical pressure, the driving force that causes electrons to move through a conductor such as a wire. Without it, electrons will not move through a wire. Voltage is produced by batteries in flashlights, solar electric modules, wind generators, and power plants.

on the outside of the house. It then flows through the wires that connect to the utility lines. From here, it travels along the power lines running by your home or business where it is consumed in neighboring homes and businesses. Once the electricity is fed onto the grid, the utility treats it as if it were theirs. End users pay the utility directly for the electricity you generate.

In most locations, an electric meter monitors the contribution of small-scale producers to the grid. The meter also keeps track of electricity the utility supplies to these homes or businesses when their PV systems aren't producing enough to meet their demands or when the PV system is not operating, for example, at night. (To learn how an electric utility measures a producer's output and how they "pay" for it, check out the sidebar, "Metering and Billing in Grid-connected Systems.")

In addition to the utility electric meter — or meters (some utilities require two or more meters) — that monitor the flow of electricity to and from the local utility grid, grid-connected solar electric systems also contain two safety disconnects. Safety disconnects are manually operated switches that enable service personnel to disconnect key points in the system to prevent electrical shock when servicing the system.

As shown in Figure 5.1b, the first disconnect is located between the solar array and the inverter. This is a DC disconnect. The manual disconnect allows the operator to terminate the flow of DC electricity from the array to the inverter in case the inverter needs to be serviced. These systems also require an AC disconnect switch. Shown in Figure 5.1b, this disconnect must be mounted outside the home or business. It must be readily accessible to utility workers and must contain a switch that can be locked in the open position by utility workers so no electricity flows to or from the grid. This disconnect is required so workers can isolate PV systems from the electrical grid and work on electrical lines without fear of shock if, for example, a line in your area goes down in an ice storm.

For many years, lockable AC disconnects were considered critical for the safety of utility personnel. Although utility-company-accessible, lockable, visible AC disconnects are required by many utilities, large California utilities with thousands of solar- and wind-electric systems now online and Colorado's main electric utility have dropped this requirement. They've found that AC disconnects are not needed because grid-compatible (synchronous) inverters automatically shut off when the utility power goes down. Properly installed PV systems will not back feed onto a dead grid. Period.

The Pros and Cons of Grid-connected Systems

Batteryless grid-connected systems represent the majority of all new solar electric systems in the United States. They're the least expensive of all systems and require the least maintenance, primarily because they contain no batteries. In essence, the electrical grid becomes your battery bank. Although popular, they do have some disadvantages, summarized in Table 5-1. You may want to take a moment to study them.

Metering and Billing in Grid-connected Systems

The idea of selling electricity to a local utility appeals to many people. But how do utilities keep track of the two-way flow of electricity from grid-connected renewable energy systems so they know how much you supply to them and how much they sell to you?

In many cases, utilities use the existing dial-type electric meter. This meter measures kilowatt-hours of energy and can run forward or backward, i.e., it is bidirectional. It runs forward when your home or business is drawing electricity from the electrical grid — for instance, if a home or business is using more electricity than its PV system is producing. It runs backward when the PV system is producing more electricity than is being used.

Some utilities install two dial-type meters, one for each direction. (Remember that utilities typically charge to install the second meter; they may also charge a separate monthly fee to read a second meter!)

In many new installations, utility companies install digital meters that keep track of the electricity delivered to and supplied by a home.

Most states have laws that require utility companies to net meter their customers. Net metering is an arrangement that ensures that customers who feed surplus electricity onto the grid can draw it back off at no charge. If, for instance, your PV system produced 100 kWh of electricity in a month that was fed onto the grid during times of excess, but drew 100 kWh of electricity off the grid, there would be no charge for the electricity they supplied. Where net metering laws vary, however, is with potential surpluses at the end of billing periods. Surpluses are known as net energy generation.

State net metering policies fall into two broad categories, annual and monthly. In annual net metering, the utility will carry surpluses, if any, up to a year. At the end ☛

of the year, they will reconcile with the customer. Monthly net metering states require utilities to reconcile net excess generation at the end of each month.

Utilities have three options when it comes to surpluses. In some states, surpluses are granted to the utility. In others, the customer is paid for surpluses at wholesale rates (what it costs the utility to generate the electricity) at the end of the billing period. In other states, utilities pay retail rates for net excess generation — that is, what they charge their customers. The best arrangement is annual net metering in which states pay retail rates.

The advantage of annual net metering is that it accommodates the seasonal variation in a PV system's production. In the summer months in many climates, a PV system will produce more electricity than is consumed by a home or business. These surpluses can be "banked" with the utility. In the winter months, when the PV system typically produces less electricity than is consumed, credits are withdrawn from the "bank" and applied to the bill.

If you're thinking that net metering could be a profitable venture, don't get your hopes up. We've only found two states — Minnesota and Wisconsin — that pay customers retail rates for monthly net excess generation. (Wisconsin only pays if the monthly net energy generation exceeds $25; otherwise the surplus is carried forward to the next month.) Most other utilities pay for surpluses at the avoided cost, that is, the cost of generating new power. (Most states reconcile monthly; nine reconcile annually.) Some states like Arkansas simply "take" the surplus without payment to the customer — it all depends on state law. (If you're not happy with your state law, you might want to consider working to change it!)

At this writing (March, 2010), 43 states, the District of Columbia and Puerto Rico have implemented statewide ☛

net metering programs, although there are substantial differences among the states. Differences include who is eligible, which utilities are required to participate in the program, the size and types of systems that qualify, and, of course, payment structure. Many states only require "investor-owned" utilities to offer net metering. In many states, municipal (city-owned) utilities and rural cooperatives are exempt from net metering laws.

Utilities that don't offer net metering may use a buy-sell or net billing system. In buy-sell schemes, utilities typically install two meters, one to track electricity the utility sells to the customer, and another to track electricity fed onto the grid by the customer. The problem with net billing arrangements is that, unlike net metering, utilities typically charge their customers retail rates for electricity used by the customer but pay customers wholesale rates for electricity fed onto the grid by a renewable energy system. For example, a utility may charge 10 to 15 cents per kilowatt-hour for electricity they supply to customers, but pay only wholesale rates of 2 to 3 cents per kilowatt-hour for energy that customers deliver to the grid. How does this work out financially for the small-scale producer?

As you might suspect, not very well. ■

The biggest downside of batteryless grid-connected PV systems is that they are vulnerable to grid failure. That is, when the grid goes down, so does the PV system. A home or business cannot use the output of a batteryless photovoltaic system when the grid is not operational. Even if the Sun is shining, batteryless grid-tied PV systems shut down if the grid experiences a problem — for instance, if a line breaks in an ice storm or lightning strikes a transformer two miles from your home or business, resulting in a power outage. Even though the Sun is shining, you'll get no power from your system.

Table 5.1 Pros and Cons of Batteryless Grid-Tied Systems	
Pros	**Cons**
Simpler than other systems	Vulnerable to grid failure unless a backup generator or an uninterruptible power supply is installed
Less expensive	
Less maintenance	
Unlimited supply of electricity (unless the grid is down)	
More efficient than battery-based systems	
Unlimited storage of surplus electricity (unless the grid is down)	
Greener than battery-based systems	

If power outages are a recurring problem in your area and you want to avoid service disruptions, you may want to consider installing an uninterruptible power supply (UPS) on critical equipment such as computers or medical equipment. A UPS has a battery pack and an inverter. If the utility power goes out, the UPS will supply power until its battery gets low. Or, you may want to consider installing a standby generator that switches on automatically when grid power goes down.

Or, as discussed in the next section, you may want to consider installing a grid-connected system with battery backup. In this case, batteries provide backup power to a home or business when the grid goes down.

Grid-connected Systems with Battery Backup

Grid-connected systems with battery backup are also known as *battery-based utility-tied systems* or *battery-based grid-connected systems*.

These systems ensure a continuous supply of electricity, despite brownouts and blackouts.

As shown in Figure 5.2, a grid-connected system with battery backup consists of (1) a PV array, (2) an inverter, (3) safety disconnects, (4) a main service panel, and (5) meters. Although grid-connected systems with battery backup are similar to battery-less grid-connected systems, they differ in several notable ways. One of the most important is the type of inverter. Battery-based grid-connected systems require a special type of inverter — one that can operate in sync with the grid, but also off grid from batteries. (We'll describe the differences in more detail in Chapter 6.)

Another more obvious difference is that battery-based grid-connected systems require a battery bank. The third difference is a meter that allows the operator to monitor the flow of electricity into and out of the battery bank. The fourth difference is the charge controller. It regulates battery charging from the PV array. (More on charge controllers in Chapter 7.)

Batteries for grid-connected systems with battery backup are either flooded lead-acid batteries or, more commonly, low-maintenance sealed lead-acid batteries. Because batteries are discussed in Chapter 7, we'll highlight only a few important considerations here.

The first point worth noting is that battery banks in grid-connected systems are typically small. That's because they are usually sized to provide sufficient storage to run a few critical loads only for a day or two while the utility company restores electrical service. Critical loads might include a few lights, the refrigerator, a well pump, the blower of a furnace, or the pump in a gas- or oil-fired boiler. It is worth noting, too, that in battery-based grid-tied systems, batteries are called into duty only when the grid goes down. They're a backup source of power; they're not there to supply additional power, for example, to run loads that exceed the PV system's output when the grid is operational. When demand exceeds supply, the grid makes up the difference, not the batteries. When the Sun is down, the grid, not the battery bank, becomes a home or business's power source.

a

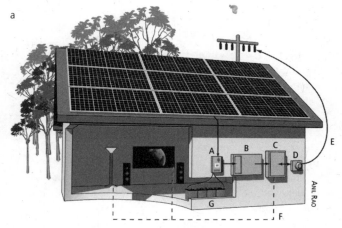

A Charge controller
B Inverter
C Breaker box
   (main service panel)
D Utility meter
E Wire to utility service
F Circuits to household loads
G Back-up battery bank

b

PV array

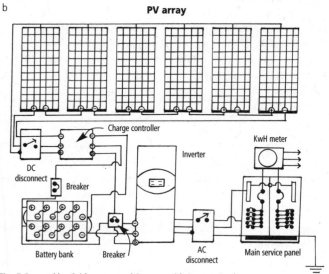

Fig. 5.2 a and b: *Grid-connected System with Battery Backup*
(a) overview and (b) drawing of system wiring and system components.

Amps and Amp-Hours

Electricity is the flow of electrons through a wire. Like water flowing through a hose, electricity flows through wires at varying rates. The rate of flow, or current, depends on the voltage. The flow of electrons through a conductor is measured in amperes, or "amps" for short. (An ampere is a certain number of electrons passing by a point per second.) The greater the amperage, the higher the number of electrons flowing through the circuit.

To gain a better understanding of power produced or power consumed, we have to consider the flow of electricity over a period of time. An *amp-hour* is the quantity of electrons flowing through a wire equal to one amp for a period of one hour. This term is also frequently used to define a battery's storage capacity. A flooded lead-acid battery, for example, might store 420 amp-hours of electricity. Because this is a pretty unfamiliar term, I recommend that amp-hours be converted to kilowatt-hours, a much more familiar term. (Individuals buy, and utility companies sell, kilowatt-hours.) To calculate kilowatt-hours from amp-hours, simply multiply the voltage of a battery by the amp-hours. A 6-volt, 420 amp-hour battery, for example, stores 2,520 watt-hours or about 2.5 kWh.

It is also important to point out that battery banks in grid-connected systems are maintained at full charge — day in and day out — to ensure a ready supply of electricity should the grid go down. Keeping batteries fully charged is a high priority of these systems. Maintaining a fully charged battery bank requires a fair amount of electricity over the long haul. This reduces system efficiency. Batteries require a continual input because they self-discharge. That is, they lose electricity when sitting idly by. You've seen it happen to flashlight batteries or car batteries that sit idle for months. Because of this, batteries require a continual electrical charge and therefore become a regular load on a renewable energy system.

In the best case, topping off batteries will consume about 5 to 10% of a system's output. In the worst case — that is, in a system with a low-efficiency, unsophisticated inverter that is used to charge a large or older battery bank — it may approach 50%.

Battery banks in grid-connected systems don't require careful monitoring like those in off-grid systems, but it is a very good idea to keep a close eye on them. When an ice storm knocks out power to your home or business, the last thing you want to discover is that your battery bank died on you last year. For this reason, grid-connected systems with battery backup typically include a meter to monitor the total amount of electricity stored in the battery bank. These meters give readings in amp-hours or kilowatt-hours. (See sidebar, "Amps and Amp-Hours" for definitions.) You'll learn more about meters to monitor batteries in Chapter 7.

Meters in battery-based systems also typically display battery voltage. For experienced renewable energy operators, battery voltage provides a general approximation of the amount of energy in a battery. If, for instance, a battery is not being charged or discharged, the higher the voltage, the more energy it holds. The lower the voltage, the less energy it stores. Because charging a battery bank raises voltage and discharging lowers voltage, you only get an accurate state-of-charge voltage when the batteries have been "at rest" for a couple of hours — that is, they have not been charged or discharged for a couple of hours.

Another component found in grid-connected systems with battery backup is the charge controller, shown in Figure 5.2. The charge controller regulates the flow of electricity into a battery bank — but only when there's a power outage. The function of charge controllers is discussed in Chapter 7. Modern charge controllers and inverters often contain a function called *maximum power point tracking* (MPPT). Discussed in more detail in Chapter 6, maximum power point tracking circuitry optimizes the output of a PV array, thus ensuring the highest possible output at all times.

Most charge controllers on the market also contain a high voltage/low voltage DC conversion function. As you will learn in

Chapter 7, this feature allows the array to be wired as high as 60 or 72 volts (nominal) and still charge a 12-, 24-, or 48-volt battery bank. This cuts down on the expense of the "home-run" wire — the wire from the array to the balance of system (BOS) components — that is, the rest of the PV system. Higher voltage results in lower current. As a result, smaller wires can be used. It also allows the array to be located farther from the BOS.

Pros and Cons of Grid-connected Systems with Battery Backup

Grid-connected systems with battery backup protect homeowners and businesses from power outages and enable them to continue to run critical loads during outages. Although battery backup may seem like a desirable feature, it does have some drawbacks, which are summarized in Table 5-2. One of the biggest downsides is higher cost. They're about 30% more expensive than batteryless grid-connected systems. The higher cost, of course, is due to the added components, especially the charge controller and the batteries. Batteries require maintenance and a warm, vented storage room or battery box, resulting in additional cost.

Table 5.2 Pros and Cons of a Battery-based Grid-Tied System	
Pros	**Cons**
Provide a reliable source of electricity	More costly than batteryless grid-connected systems
Provide emergency power during a utility outage	Less efficient than batteryless grid-connected systems
	Less environmentally friendly than batteryless systems
	Require more maintenance than batteryless grid-connected systems

Because grid-connected systems with battery backup are expensive and infrequently required, few people install them. When contemplating a battery-based grid-tied system, you need to answer three questions: (1) How frequently does the grid fail in your area? (2) What critical loads are present and how important is it to keep them running? (3) How do you react when the grid fails?

If the local grid is extremely reliable, you don't have medical support equipment to run, your computers aren't needed for business or financial transactions, you have a wood stove for heat and you don't mind using candles on the rare occasions when the grid goes down, why buy, maintain, and replace costly batteries?

In some cases, people are willing to pay for the reliability that a battery bank brings to a grid-connected system.

Off-Grid Systems

Off-grid systems are designed for individuals and businesses that want to or must supply all of their needs via solar energy — or a combination of solar and wind or some other renewable source. As shown in Figure 5.3a, off-grid systems bear a remarkable resemblance to grid-connected systems with battery backup. There are some noteworthy differences, however. The most notable is the lack of grid connection.

As illustrated in Figure 5.3, electricity flows from the PV array to the charge controller. The charge controller monitors battery voltage and delivers DC electricity to the battery bank. When electricity is needed in a home or business, it is drawn from the battery bank via the inverter. The inverter converts the DC electricity from the battery bank, typically 24 or 48 volts in a standard system, to higher-voltage AC, either 120 or 240 volts, which is required by households and businesses. AC electricity then flows to active circuits in the house via the main service panel.

Off-grid systems often require a little "assistance," to make up for shortfalls. Additional electricity can be supplied by a wind turbine, micro hydro turbine, or a gasoline or diesel generator, often referred to as a gen-set. "A gen-set also provides redundancy," notes

a

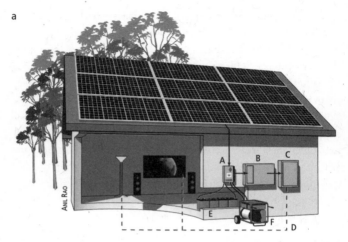

A Charge controller	D Circuits to household loads
B Inverter	E Battery bank
C Breaker box	F Back-up generator

b

PV array

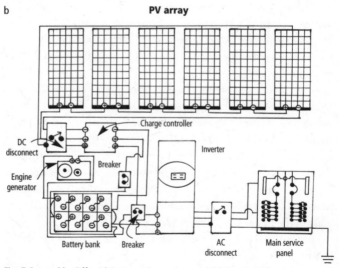

Fig. 5.3 a and b: *Off-grid System. a) overview and (b) drawing of system wiring and system components.*

National Renewable Energy Laboratory's wind energy expert Jim Green. Moreover, "if a critical component of a hybrid system goes down temporarily, the gen-set can fill in while repairs are made." Gen-sets also play a key role in maintaining batteries, a subject discussed in Chapter 7.

Off-grid systems with gen-sets require another component, a battery charger. They convert the AC electricity produced by the generator into DC electricity that's then fed into the battery bank. Battery chargers are built into the inverter and operate automatically. When a generator is started and the inverter senses voltage at its input terminals, it then transfers the home loads over to the generator through an internal, automatic transfer switch. It also begins charging the battery from the generator.

Like grid-connected systems with battery backup, an off-grid system requires safety disconnects — to permit safe servicing. DC disconnects, with appropriately rated fuses or breakers, are located between the PV array and the charge controller, between the charge controller and the battery bank, and between the battery and the inverter.

These systems also require charge controllers to regulate battery charging from the PV array. Charge controllers also protect the batteries from overcharging.

As is evident by comparing schematics of the three types of systems, off-grid PV systems are the most complex. Moreover, some systems are partially wired for DC — that is, they contain DC circuits. These are supplied directly from the battery bank. DC circuits are used to service lights or DC appliances such as refrigerators or DC well or cistern pumps. Why include DC circuits?

Many people who install them do so because DC circuits bypass the inverter. Because inverters are not 100% efficient in their conversion of DC to AC, this saves energy. Operating a DC refrigerator, for example, over long periods can save a substantial amount of energy.

The problem with this strategy is that DC circuits are low voltage circuits and thus require much larger wiring and special, more

Fig. 5.4: *Power Center. Power centers like the one shown here contain all the key components of a PV system, including one or more inverters, charge controllers, fuses and disconnects.*

expensive plugs and sockets. DC appliances are also harder to find. In addition, they are typically much smaller than those used in homes, and they are less reliable.

If you are thinking about installing an off-grid system in a home or business, your best bet is an AC system — unless your home is extremely small and your needs are few.

To simplify installation of battery-based systems, many installers recommend use of a power center, such as the one shown in Figure 5.4. Power centers contain many of the essential components of a renewable energy system, including the inverter, the charge controller, and fused safety disconnects — all prewired. This makes an electrician's job easier. Power centers also provide busses (connection points) to which the wires leading to the battery bank, the inverter, and the PV array connect.

Pros and Cons of Off-Grid Systems

Off-grid systems offer many benefits, including total emancipation from the electric utility (Table 5-3). They provide a high degree of energy independence that many people long for. You become your own utility, responsible for all of your energy production. In addition, if designed and operated correctly, your system will provide energy day in and day out for many years. Off-grid systems also provide freedom from occasional power failures.

Table 5.3 Pros and Cons of an Off-grid System	
Pros	**Cons**
Provide a reliable source of electricity	Generally the most costly solar electric systems
Provide freedom from utility grid	Less efficient than batteryless grid-connected systems
Can be cheaper to install than grid-connected systems if located more than 0.2 miles from grid	Require more maintenance than batteryless grid-connected systems (you take on all of the utility operation and maintenance jobs and costs)

Off-grid systems do have some downsides, summarized in Table 5-3. One of the most significant is that they are the most expensive of the three renewable energy system options. Battery banks, supplemental wind systems, and generators add substantially to the cost — often 60% more. They also require more wiring. In addition, you will need space to house battery banks and generators. Although cost is usually a major downside, there are times when off-grid systems cost the same or less than grid-connected systems — for example, if a home or business is located more than a few tenths of a mile from the electric grid. Under such circumstances, it can cost more to run electric lines to a home than to install an off-grid system.

When installing an off-grid system, remember that you become the local power company and your independence comes at a cost to you. Also, although you may be "independent" from the utility, you will need to buy a gen-set and fuel, both from large corporations. Gen-sets cost money to maintain and operate. You will be dependent on your own ability to repair your power system when something fails.

An off-grid system also comes at a cost to the environment. Gen-sets produce air and noise pollution. Lead-acid batteries are far from environmentally benign. Although virtually all lead-acid batteries are recycled, battery production is responsible for considerable environmental degradation. Mining and refining the lead are fairly damaging. Thanks to NAFTA and the global economy, lead production and battery recycling are being carried out in many poor countries that have lax or nonexistent environmental policies. They are responsible for some of the most egregious pollution and health problems facing poorer nations across the globe, according to small wind energy expert Mick Sagrillo. So, think carefully before you decide to install an off-grid system.

Hybrid Systems

All three solar systems can be designed to include one or more additional renewable energy sources. The result is a hybrid renewable energy system (Figure 5.5).

Hybrid renewable energy systems are extremely popular among homeowners in rural areas. Solar electricity and wind are a marriage made in heaven in many parts of the world. Why?

In most locations, solar energy and winds vary throughout the year. Solar radiation striking the Earth tends to be highest in the spring, summer, and early fall. Winds tend to be strongest in the late fall, winter, and early spring.

Table 5-4 shows that sunlight is relatively abundant in central Missouri from March through October. Table 5-5 shows that winds, however, pick up in October and blow through May.

Table 5.4
Solar Resource near Gerald, Missouri measured in kWh/m² per day*

Lat 38.325 Lon -91.297	Jan	Feb	Mar	Apr	May	Jun
Tilt 38	3.66	3.86	4.71	5.43	5.28	5.50

*Note: This data represents solar energy striking a collector mounted at an optimal angle for this location.

Together, solar electricity and wind can provide 100% of a family's or business' electrical energy needs. This complementary relationship is shown graphically in Figure 5.6.

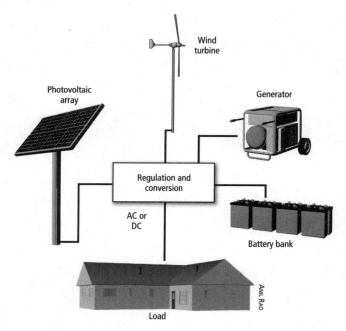

Fig. 5.5: *Hybrid PV and Wind System*

Jul	Aug	Sep	Oct	Nov	Dec	Annual Average
5.75	5.66	5.48	4.79	3.27	2.95	4.70

Table 5.5
Ten-Year Monthly Average Wind Speed
Near Gerald, Missouri at 120 feet

Lat 38.325 Lon -91.297	Jan	Feb	Mar	Apr	May	Jun
m/s	5.96	5.93	6.50	6.31	5.27	4.83
m/hr	13.33	13.26	14.54	14.12	11.79	10.8

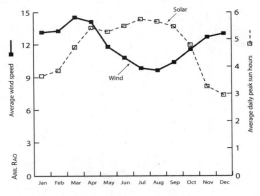

Fig. 5.6: Complementary nature of wind and solar near Gerald, Missouri.

In areas with a sufficient solar and wind resource, a properly sized hybrid PV/wind system can not only provide 100% of your electricity, it may eliminate the need for a backup generator. Because wind and PVs complement each another, you can install a smaller solar electric array and a smaller wind generator than if either were the sole source of electricity.

Hybrid systems may also make sense for those installing a grid-tied system in areas with net metering laws that stipulate a low sell price; a hybrid system reduces the amount of electricity that must be bought. For grid-tied systems with annual net metering or buy-sell with a sell price equal to or higher than the buy price, reducing seasonal variation in production has no economic advantages. The best strategy is to make as much electricity as you can at the lowest

Jul	Aug	Sep	Oct	Nov	Dec	Annual Average
4.40	4.31	4.64	5.16	5.67	5.82	5.40
9.84	9.64	10.38	11.54	12.68	13.02	12.08

cost. In this case, only one technology makes economic sense, the one with the lowest cost.

If the combined solar and wind resource is not sufficient throughout the year or the system is undersized, a hybrid system will require a backup generator — a gen-set to supply electricity during periods of low wind and low sunshine. Gen-sets are also used to maintain batteries in peak condition (discussed in Chapter 7), and permit use of a smaller battery bank.

Choosing a PV System

To sum things up, homeowners and businesses have three basic choices when it comes to installing a PV system. If you have access to the electric grid, you can install a batteryless grid-connected system, by far the cheapest and simplest option. Or you can install a grid-connected system with battery backup. If you don't have access to the grid, you can install an off-grid system. All of these systems can combine two or more sources of electricity, creating hybrid systems.

If you are building a home close to a utility line, a grid-connected system is often a good choice. This system will allow you to use the grid to store excess electricity. Although you may encounter occasional power outages, in most places these are rare and transient occurrences.

If you are installing a PV system on an existing home or business that is already connected to the grid, it is generally best to stay connected. Use the grid as your battery bank. Grid-connected systems with battery banks are suitable for those who want to stay connected to the grid, but also want to protect themselves from occasional blackouts. They'll cost more, but they provide peace of mind and security. It's best to back up only the truly critical loads and make sure they are highly efficient. Doing so will reduce the size of the battery required to meet the loads, reduce the cost of the system, and improve the efficiency of the system. Some loads, like a forced air furnace, can be quite a challenge to back up. Not only is a furnace blower one of the larger loads in a home, it also runs during the time of the year with the least amount of sunlight.

Although more expensive than grid-connected systems, off-grid systems are often the system of choice for customers in remote rural locations. When building a new home in a rural location, grid connection can be pricey — so pricey that an off-grid system makes good sense. Some utility companies, however, pay for line extension, connection, and metering. Be sure to check, when considering which system you should install. "Utility policies vary considerably when it comes to line extension costs," notes NREL's Jim Green. "Sometimes, the utility absorbs much of the cost in the rate base. Others pass most or all of the cost to the new customer."

CHAPTER 6

INVERTERS

The inverter is an indispensable component of virtually all electric-generating renewable energy systems. In this chapter, I'll discuss the types of inverters and the functions they provide. (For a detailed look inside an inverter to see how it operates, you may want to pick up a copy of my book *Power from the Sun*.)

Types of Inverters

Inverters come in three basic types: those designed for grid-connected systems, those made for off-grid systems, and those designed for grid-connected systems with battery backup.

Grid-Connected Inverters

Today, the vast majority of renewable energy systems — both solar and wind electric — are grid-connected. These systems require inverters that operate in sync with the utility grid. They produce electricity that is identical to that available on the grid. This type of electricity is known as *grid-compatible sine wave AC electricity*.

Grid-connected inverters are also known as *utility-tie inverters*. They convert DC electricity from the array in a PV system into AC electricity (Figure 6.1). Electricity then flows from the inverter to the breaker box and is then fed into active circuits, powering refrigerators, computers, and the like. Surplus electricity is then back-fed onto the grid, running the electrical meter backward.

Fig. 6.1: *Grid-Connected Inverter.*
This inverter by Fronius is designed
for batteryless grid-connected wind
energy systems.

Grid-tied inverters produce electricity that matches the grid both in frequency and voltage. To do this, these inverters monitor the voltage and frequency of the electricity on the utility lines. They then adjust their output so that it matches that available on the grid. That way, electricity that is fed from a PV system onto utility lines is identical to the electricity utilities are transmitting to their customers.

Grid-compatible inverters are equipped with *anti-islanding protection* — a feature that automatically disconnects the inverter from the grid in case of loss of grid power. That is, grid-connected inverters are programmed to shut down if the grid goes down. The inverter stays off until service is restored. This feature protects utility workers from electrical shock.

Grid-compatible inverters also shut down if there's an increase or decrease in either the frequency or voltage of grid power outside the inverter's acceptable limits (established by the utility companies). If either varies from the inverter's pre-programmed settings, the inverter turns off.

The inverter shuts down entirely in the case of blackouts, because it requires grid connection to determine the frequency and voltage of the AC electricity it produces. Without the connection, the inverter can't operate.

Grid-connected inverters also come with a *fault condition reset* —
a sensor and a switch that turns the inverter on when the grid is
back up or the inverter senses the proper voltage and/or frequency.

To avoid losing power when the grid goes down, a homeowner
can install a grid-connected system with battery backup. Although
inverters in such systems disconnect from the utility during outages, they can draw electricity from the battery bank to supply
active circuits. As noted in the previous chapters, such systems are
typically designed and wired to provide electricity only to essential
circuits in a home or business, supplying the most important (critical) loads.

Grid-connected inverters also frequently contain LCD displays
that provide information on the input voltage (the voltage of the
electricity from the PV array) and the output voltage (the voltage
of the AC electricity the inverter produces and delivers to a home
and the grid). They also display the current (amps) of the AC output.

Some inverters, like the Trace SW utility-intertie series inverters, come with automatic morning wake-up and evening shutdown.
These features shut the inverter down at night (as it is no longer
needed) and wake it up in the morning (to get ready to start converting DC electricity from the array into AC electricity.) This sleep
mode in the SW series inverters uses less than 1 watt of power.

Grid-connected inverters operate with a fairly wide input
range. The DC operating range of the SW series, for instance,
ranges from 34 to 75 volts DC (you might see this listed as
"VDC"). Inverters from Fronius and Oregon-based PV Powered
Design are designed to operate within a broader DC voltage input
range: from 150 to 500 volts DC. This permits the use of a wider
range of modules and system configurations. Moreover, high-voltage arrays can be placed farther from the inverter than low-voltage
arrays. In addition, high-voltage DC input means that smaller and
less expensive wires can be used to transmit electricity to a home or
office from the array. With the cost of copper skyrocketing as a
result of higher energy prices and higher demand, savings on wire
size can be substantial.

Off-Grid Inverters

Like grid-connected inverters, off-grid inverters convert DC electricity into AC and boost the voltage to 120 or 240 volts. Off-grid inverters also perform a number of other essential functions, discussed here. If you're installing an off-grid system, be sure to read this section carefully.

Battery-based inverters used in off-grid and grid-connected systems with battery backup typically contain battery chargers. Battery chargers charge batteries from an external source — usually a genset in an off-grid system. But isn't the battery charged by the PV array through the charge controller?

The charge controller in battery-based systems does indeed charge batteries, however, its job is to charge batteries from the PV array, not a gen-set. The charge controller therefore receives DC electricity from a wind turbine PV array, then sends it to the battery bank. The charge controller also prevents batteries from being overcharged. A battery charger in the inverter, on the other hand, converts AC from a gen-set and converts it to DC. It then feeds DC electricity to the batteries.

In off-grid systems, battery charging gen-sets are used to restore battery charge after periods of deep discharge. As noted in Chapter 7, this prolongs battery life and prevents irreparable damage to the plates. Battery chargers are also used during equalization, also discussed in Chapter 7.

High-quality battery-based inverters also contain programmable high- and low-voltage disconnects. These protect various components of the system, such as the batteries, appliances, and electronics in a home or business. They also protect the inverters.

The high-voltage disconnect is a sensor/switch that terminates the flow of electricity from the gen-set when the batteries are charging if the battery voltage is extremely high. (Remember: high battery voltages indicate that the batteries are full.) High-voltage protection therefore prevents overcharging, which can severely damage the lead plates in batteries. It also protects the inverter from excessive battery voltage. The low-voltage disconnect (LVD) in an inverter monitors battery voltage at all times. When low battery voltage is detected (indicating the batteries are deeply discharged) the inverter shuts off and often sounds an alarm. The flow of electricity from the batteries to the inverter stops. The inverter stays off until the batteries are recharged.

Low-voltage disconnect features are designed to protect batteries from very deep discharging. Although lead-acid batteries are designed to withstand deep discharges, discharging batteries beyond the 80% mark causes irreparable damage to the lead plates in batteries and leads to their early demise. Although complete system shutdown can be a nuisance, it is vital to the survival of a battery bank.

Batteries can be recharged by a supplementary wind turbine or by a gen-set. Gen-sets may be manually started, although some inverters contain a sensor and switch that activates the generator automatically when low battery voltage is detected. The fossil fuel generator then recharges the batteries using the inverter's battery charger.

Multifunction Inverters

Grid-connected systems with battery backup require multifunction inverters — battery- and grid-compatible sine wave inverters (Figure 6.2). They're commonly referred to as *multifunction* or, less commonly, *multimode inverters*.

Multifunction inverters contain features of grid-connected and off-grid inverters. Like a grid-connected inverter, they contain anti-islanding protection that automatically disconnects the inverter from the grid in case of loss of grid power, over/under voltage, or over/under frequency. They also contain a fault condition reset to power up an inverter when a problem with the utility grid is corrected. Like off-grid inverters, multifunction inverters contain battery chargers and high- and low-voltage disconnects.

If you are installing an off-grid system, you may want to consider installing a multifunction inverter in case you decide to connect to the grid in the future. However, even though multifunction inverters allow system flexibility, they are not always the

XANTREX (XW SERIES INVERTER)

Fig. 6.2: *Multifunction Inverter. This inverter from Xantrex can be used for grid-connected systems with batteries and off-grid systems.*

most efficient inverters. That's because some portion of the electricity generated in such a system must be used to keep the batteries topped off. This may only require a few percent, but over time a few percent add up. In systems with poorly designed inverters or large backup battery banks, the electricity required to maintain the batteries can be quite substantial. It is also worth noting that as batteries age, they become less efficient; more power is consumed just to maintain the float charge, that is, a constant "resting" voltage. This reduces the efficiency of the entire PV system.

For best results, I recommend inverters that prioritize the delivery of surplus electricity to the grid while preventing deep discharge of the battery bank, such as OutBack's multifunction inverters and Xantrex's XW series inverters.

If you want the security of battery backup in a grid-connected system, I suggest that you isolate and power only critical loads from the battery bank. This minimizes the size of your battery bank, reduces system losses, and reduces costs. Unless you suffer frequent or sustained utility outages, a batteryless grid-connected system usually makes more sense from both an economic and environmental perspective.

Buying an Inverter

Inverters come in many shapes, sizes, and prices. Most homes and small businesses require inverters in the 2,500 to 5,500-watt range. Which inverter should you select?

When purchasing an inverter through an installer, your choices may be limited. Most installers carry a line of inverters they are familiar with and have a high degree of confidence in. If this is the case, the installer will make a recommendation that fits your needs from the inverters he or she carries.

System Voltage

When shopping for a battery-based inverter, you'll need to select one with an input and output voltage that corresponds with the input voltage of the battery bank in your system. System voltage refers to the voltage of the electricity produced by a renewable energy technology — a wind turbine, a solar electric array, or micro hydro generator. In PV systems, the arrays are wired to produce either 12-, 24-, or 48-volt electricity (although this is changing for reasons discussed in the accompanying box). The batteries are wired similarly.

Because all components of an off-grid renewable energy system must operate at the same voltage, the inverter must match the source (PV array) and batteries (if any). A 24-volt inverter won't work in a 48-volt system. If you are installing a 48-volt battery bank, you'll need a 48-volt battery-based inverter.

Modified Square Wave vs. Sine Wave

The next inverter selection criterion is the output waveform — basically, the type of AC electricity they produce. (The output wave is a graphical representation of the voltage, a topic discussed in

PV Array Voltage, String Size, and Choosing the Correct Inverter

In the 1970s through the mid 1990s, most PV systems were designed to charge 12-, 24-, or 48-volt battery banks installed in off-grid systems, the predominant type of PV system being installed then. To meet the demands of this market, manufacturers produced modules rated at 12 volts and installers built systems around them. In a 12-volt system, for instance, an installer would wire one or more 12-volt modules in parallel to produce the voltage needed to charge a 12-volt battery bank. In 48-volt systems, installers would wire four 12-volt modules in series, then install one or more strings (a string is a group of modules wired in series) in parallel to produce 48-volt DC electricity.

In these systems, the 12-, 24-, or 48-volt DC electricity was then fed to a 12-, 24-, or 48-volt charge controller, which delivered the electricity to appropriately sized battery banks. Appropriate inverters were installed to draw off the 12-, 24, or 48-volt battery banks. They converted the low-voltage DC electricity into 120-volt AC electricity suitable for household use.

Today, systems based on 12-volt increments have largely fallen by the wayside. One reason for this is that most modern PV systems are batteryless grid-connected. Because there are no batteries, the inverters used in these systems can be designed to accept DC electricity from arrays at much higher voltages. In fact, most grid-connected inverters operate with an input range of 150 to 550 volts with a maximum of 600 volts. (The National Electrical Code prohibits wiring PV arrays for homes higher than 600 volts.)

Higher voltage reduces losses as electricity flows from the array to the inverter. Because less current flows at higher voltage, higher voltage systems permit the use of smaller gauge conductors, which reduces installation costs. The higher voltage also allows smaller components in the inverters, reducing their size, weight, and cost. ☛

To accommodate the higher voltage inverters, manufacturers now produce PV modules with higher nominal voltages — 16, 24, and 36 volts. These modules are then wired in series strings to produce high-voltage DC electricity for today's high-voltage inverters.

To help installers determine how many modules they can wire in a string, virtually all inverter manufacturers provide online calculators. To calculate string size, the installer simply enters the temperature conditions of the site and the type of module he or she is considering. The online calculator provides the maximum, minimum, and ideal number of modules in a series string for each of its inverters.

Temperature is important when sizing an array because output changes with temperature. Low temperatures, for example, increase the output of an array. If an array has not been sized carefully, the voltage of the incoming electricity could exceed the rated capacity of the inverter and the charge controllers — if it is a battery-based system. High voltage can damage this equipment. (This is most likely to occur early on cold, sunny days when the voltage is highest.)

High temperatures, in contrast, reduce the output of an array. If the voltage of a PV array falls below the rated input of a direct grid-tie inverter, a PV system will shut down and will remain off until the array cools down and the voltage increases. This, of course, reduces the efficiency of a system.

Many modern battery-based systems can be designed to handle higher voltage arrays thanks to the introduction of maximum power point tracking (MPPT) charge controllers. MPPT charge controllers optimize power output of an array and accept higher voltage input from PV arrays. That's because they contain step-down transformers that reduce the high-voltage array input to charge the battery bank at the appropriate voltage. These systems are generally designed with an open-circuit voltage of around 150 volts, and can be used to charge 12-, 24-, or 48-volt battery banks. (Open circuit voltage is the voltage that is generated by a module with ☞

no amps flowing, usually first thing in the morning, when there's just not enough sunlight to produce useful current flow through the charge controller.)

A PV array's operating voltage is influenced by the ambient temperature, as just noted, and also the number of modules. Array temperature is also influenced by the type of mounting. As you'll learn in Chapter 8, some array mounts allow more air to circulate around modules than others. Other mounting options place PV modules in arrays very close to hot roof surfaces. The hotter the array, the greater the decline in its output. When calculating string size, then, an installer must also stipulate the type of mounting.

Designers must also take into account the slow degradation of modules, which reduces their output over time. (Module output may decline as a result of moisture that seeps into the PV cells, which corrodes the internal electrical connections. Module deterioration also results from the slow deterioration of the thin, clear plastic coating on the PV cells.) Because of this, it is important *not* to design a system that operates consistently near the lower end of the inverter's input voltage window. As the modules' output declines over time, you'll increase the chances of reduced energy output on hot days. ■

much more detail in *Power from the Sun.*) Battery-based inverters are available in modified square wave (often called modified sine wave) and sine wave. Grid-connected inverters are all sine wave so their output matches utility power.

Modified square wave electricity is a crude approximation of grid power that works fairly well in many appliances and electrical devices in our homes and businesses (Figure 6.3). Although most office and household electronic equipment and appliances can function on modified square wave electricity, they all run less efficiently, producing less of what you want — i.e., light, pumped water, etc. — and more waste heat for a given energy input. When operated on

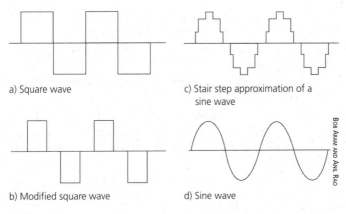

a) Square wave

c) Stair step approximation of a
sine wave

b) Modified square wave

d) Sine wave

Fig. 6.3

modified square wave electricity, microwave ovens cook slower. Equipment and appliances that run warmer might last fewer years on modified square wave electricity. Computers and other digital devices operate with more errors and crashes. Digital clocks don't maintain their settings as well. Modified square wave electricity may cause an annoying, high-pitched buzz or hum on TVs and stereos and may also produce annoying lines on TV sets. It can even damage sensitive electronic equipment. Some equipment, like modern washing machines, may not operate at all on modified square wave electricity. The computer that controls them won't run on it. Unless money is tight, I recommend sine wave battery-based inverters for off-grid systems. SMA, Xantrex, and OutBack all produce excellent sine wave inverters at reasonable prices.

Output Power, Surge Capacity, and Efficiency

When selecting an inverter for an off-grid system, be sure to pay attention to continuous output, surge capacity, and efficiency.

Continuous Output

Continuous output is the power an inverter can produce on a continuous basis. It is measured in watts, although some inverter spec sheets also list continuous output in amps (to convert, simply use the formula watts = amps x volts). OutBack's sine wave inverter VFX3524 produces 3,500 watts of continuous power and is designed for use in 24-volt systems. The 35 in the model number stands for 3,500 watts. The 24 indicates it is designed for a 24-volt system.

To determine the continuous output you'll need, add up the wattages of the common appliances you think will be operating at once. Be reasonable, though. Typically, only two or three large loads operate simultaneously.

Surge Capacity

Electrical devices with motors, such as vacuum cleaners, refrigerators, washing machines, and power tools, require a surge of power to start up. It typically lasts only a fraction of a second. Even though the power surge is brief, if an inverter can't provide the power, the motor won't start. Moreover, the stalled motor will draw excessive current and could overheat, unless it is protected by a thermal cutout. If not, it may burn out.

When shopping for an inverter, then, be sure to check out the surge capacity. All quality inverters are designed to permit a large surge of power over a short period, usually around five seconds. Surge power exceeds the continuous output and is listed on spec sheets in either watts or amps.

Efficiency

Converting one form of energy to another results in a loss of energy. Efficiency is calculated by dividing the energy coming out by the energy going in. Fortunately, efficiency losses in inverters are quite low — usually only 5 to at the most 15 percent. It should be noted, however, that inverter efficiency varies with load. Generally, an inverter doesn't achieve its highest efficiency until output reaches 20 to 30% of its rated capacity, according to Richard Perez, author, renewable energy expert, and publisher of *Home Power* magazine. A 4,000-watt inverter, for instance, will be most efficient at outputs of

800 to 1,200 watts. At lower outputs, efficiency is dramatically reduced.

To get the most out of a PV array, most systems incorporate a function known as *maximum power point tracking* (MPPT). The circuitry that allows MPPT is located in the inverter in grid-connected systems and in the charge controller of battery-based systems. Maximum power point tracking ensures that the array produces the most power it can during daylight hours. The details of how MPPT works is beyond the scope of this book. Those wishing to learn more may want to check out *Power from the Sun*.

Noise and Other Considerations

Battery-based inverters are typically installed inside, close to the batteries to reduce line loss. Grid-tied inverters are almost always installed near the main circuit breaker panel where the utility service comes into the house. (Most inverter manufacturers like their equipment to be housed at room temperature.)

If you are planning to install an inverter inside your home or office, be sure to check out the noise it produces. Inquire about this upfront; better yet, ask to listen to the model you are considering in operation.

Some folks are concerned about the potential health effects of extremely low frequency electromagnetic waves emitted by inverters as well as other electronic equipment and electrical wires. If you are concerned about this, install your inverter in a place away from people. Avoid locations in which people will be spending a lot of time — for example, don't install the inverter on the other side of a wall from your bedroom or office.

While you're developing a checklist of features to consider when purchasing an inverter, be sure to add ease of programming. Find out in advance how easy it is to change settings. Spend some time with the manual.

Stackability

Finally, when buying a battery-based inverter, you may also want to select one that can be stacked — that is, connected to a second or third inverter of the same kind. Stacking permits homeowners to produce more electricity in case demands increase over time. Two inverters can be wired in parallel, for example, to double the amp output of a battery-based PV system.

A couple of inverter manufacturers such as OutBack and Xantrex produce power panels, easily mounted assemblies that house two or more inverters for stacking. Figure 6.4 shows a power panel by OutBack that contains four inverters.

Many homes require 240-volt electricity to operate appliances such as electric clothes dryers, electric stoves, central air conditioning, or electric resistance heat. We recommend that you avoid such appliances, especially when installing an off-grid system. That's not because a PV or hybrid PV and wind system can't meet those demands, but rather because these appliances use lots of electricity, so you'll need a larger and more costly PV system to power them. Well-designed, energy-efficient homes can usually avoid using 240 VAC. An exception is a deep well pump, which may require 240-volt electricity. In most cases, high efficiency 120-volt AC pumps, or even DC pumps, perform admirably.

Fig. 6.4: *OutBack Power Panel. Power panels like this one contain all the components needed for a PV system. This one has four inverters and is for a very large PV system.*

If you must have 240-volt AC electricity, purchase an inverter that can be wired in series to produce 240 VAC, such as the OutBack FX series inverters. Or you can purchase an inverter that produces 120- and 240-volt electricity, such as Magnum Energy's MS-AE 120/240V Series Inverter/Charger or one of Xantrex's new XW series inverters. Or you can install a step-up transformer that converts 120-volt AC electricity from the inverter to 240-volt AC. Or you can simply install a dedicated 240-volt output inverter for that load.

Conclusion

A good inverter is key to the success of a solar energy system, so shop carefully. Size it appropriately. Be sure to consider future electrical needs. But don't forget that you can trim electrical consumption by installing efficient electronic devices and appliances. Efficiency is always cheaper than adding more capacity! When shopping, select the features you want and buy the best inverter you can afford. Although modified square wave inverters work for most applications, you will most likely be happier with a sine wave inverter.

CHAPTER 7

BATTERIES, CHARGE
CONTROLLERS, AND GEN-SETS

If you are installing an off-grid system or a grid-connected system with battery backup, you'll need a high-quality battery bank. Because batteries are so important to the success of these systems and require so much attention, it is essential that you understand your options, how batteries work, and how they should be installed and maintained. It is also important to understand charge controllers and gen-sets. This chapter will cover all three of these important topics.

Flooded Lead Acid Batteries

Batteries used in most off-grid renewable energy systems are deep-cycle flooded lead-acid batteries. These batteries can be charged and discharged (cycled) hundreds of times before they wear out.

Lead-acid batteries contain three separate 2-volt compartments, known as cells. Inside each cell is a series of thick parallel lead plates (Figure 7.1). The cells are connected internally (wired in series) so that they produce 6-volt electricity. The space between the plates is filled with sulfuric acid (hence the term "flooded"). A partition wall separates each cell, so that fluid cannot flow from one cell to the next. The cells are encased in a heavy-duty plastic case.

As illustrated in Figure 7.1, lead acid batteries contain two types of plates: positive and negative. The positive plates connect to a positive metal post or terminal; the negative plates connect to a negative post. The posts allow electricity to flow in and out of batteries.

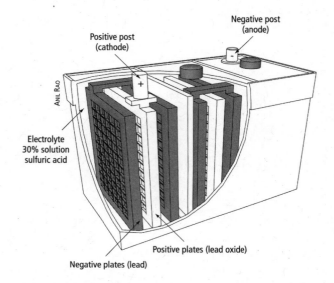

Fig. 7.1: *Anatomy of a Flooded Lead-Acid Battery*

The positive plates of lead-acid batteries are made from lead dioxide (PbO_2). The negative plates are made from pure lead. The sulfuric acid that fills the spaces between the plates is referred to as the *electrolyte*.

How Lead-Acid Batteries Work

Like all other types of batteries, lead-acid batteries convert electrical energy into chemical energy when they are charged. When discharging, that is, giving off electricity, chemical energy is converted back into electricity. The chemical reactions that take place during battery discharge are shown in Figure 7.2.

As illustrated, when electricity is drawn from a lead-acid battery, sulfuric acid reacts with the lead of the negative plates (top reaction). This reaction yields electrons, tiny negatively charged particles. They flow out of the battery creating an electrical current. During this reaction, lead on the surface of the negative plates is converted to tiny lead sulfate crystals.

negative plates

$$Pb(s) + HSO_4^-(aq) \longrightarrow PbSO_4(s) + H^+(aq) + 2e^-$$

positive plates

$$PbO_2(s) + HSO_4^-(aq) + 3H^+(aq) + 2e^- \longrightarrow PbSO_4(s) + 2H_2O$$

Fig. 7.2: *Chemical Reactions in a Lead-Acid Battery.*

When a battery is discharging, sulfuric acid also reacts with the lead dioxide of the positive plates, resulting in the formation of lead sulfate crystals on them as well (see bottom panel). Discharging a battery not only creates lead sulfate crystals on the positive and negative plates, it depletes the amount of sulfuric acid in the battery. When the battery is charged, however, lead sulfate crystals on the positive and negative plates are broken down, releasing sulfate ions into solution, thus replenishing the sulfuric acid. (The reactions that take place during recharge are the reverse of those that occur during discharge.)

Although the chemistry of lead-acid batteries is a bit complicated, it is important to remember that this system works because electrons can be stored in the chemicals within the battery when a battery is charged. The stored electrons can be drawn out by reversing the chemical reactions. Through this reversible chemical reaction, the battery is acting as a "charge pump," moving electrical charges through a circuit on demand.

Will Any Lead-Acid Battery Work?

Lead-acid batteries come in many varieties, each one designed for a specific application. Car batteries, for example, are designed and manufactured for use in cars, light trucks, and vans. Marine batteries are designed for boats; golf cart batteries are designed for use in golf carts. Forklift batteries are used to power electric forklifts.

For off-grid systems, you have three options: (1) deep-cycle flooded lead-acid battery like those made by Trojan, Rolls, Deka and others (Figure 7.3); (2) forklift batteries; and (3) golf cart batteries.

Car batteries won't work. Their lead plates are much too thin to withstand deep cycling (deep discharges), which commonly occurs in renewable energy systems. Although lead sulfate crystals that form on the plates of a battery during deep discharge are removed when these batteries are recharged, some crystals fall off before recharge occurs. The thin plates of a car battery would be whittled away to nothing in a very short time. After twenty or so deep discharges, the batteries would be ruined — no longer able to accept a charge.

Renewable energy systems require deep-discharge lead-acid batteries with thick lead plates. The thickness of the lead plates allows them to withstand multiple deep discharges — so long as they're recharged soon afterward. Even though the plates lose lead over time, they are so thick that the small losses are insignificant. As a result, a deep-cycle battery can be deeply discharged hundreds, sometimes a few thousand times, over its lifetime.

For optimum long-term performance, however, deep-cycle batteries still need to be recharged promptly after deep discharging.

Fig. 7.3: *Deep-Cycle Lead-Acid Batteries. These batteries contain thick lead plates and are used in many battery-based renewable energy systems. The thick plates permit deep cycling so long as the batteries are recharged soon after each deep discharge.*

SURRETTE BATTERY COMPANY LTD.

Don't ever forget this! With proper care, good, high capacity deep-cycle batteries like those most often used in PV systems can last seven to ten years, maybe even longer.

Forklift batteries are high-capacity, deep-discharge batteries designed for a fairly long life, and they operate under fairly demanding conditions. They can withstand 1,000 to 2,000 deep discharges — more than many other deep-cycle batteries used in battery-based renewable energy systems — and they work well in renewable energy systems, too. They are, however, rather heavy, bulky, and expensive. But, if you can acquire them at a decent price, you may want to use them.

Golf cart batteries may also work. Like forklift batteries, golf cart batteries are designed for deep discharge. However, they typically cost a lot less than other heavier duty deep-cycle batteries. While the lower cost may be appealing, golf cart batteries don't last as long as the alternatives. They may last only five to seven years, if well cared for. Shorter lifespan means more frequent replacement. More frequent replacement means higher long-term costs and more hassle.

What about Used Batteries?

Another option for cost-conscious home and business owners is a used battery. Although used batteries can often be purchased at bargain prices, they're rarely worth it. Used batteries are often discarded because they've failed or have experienced a serious decline in function. As a buyer, you have no idea how well — or how poorly — they've been treated. Have they been deeply discharged many times? Have they been left in a state of deep discharge for long periods? Have they been filled with tap water rather than distilled water? Although there are exceptions, most people who've purchased used flooded lead-acid batteries have been disappointed.

When shopping for batteries for a renewable energy system, I recommend that you buy high quality deep-cycle batteries. Although you might be able to save some money by purchasing cheaper alternatives, including used batteries, frequent replacement is time consuming. Batteries are heavy and it takes quite a lot of time to disconnect old batteries and rewire new ones. Bottom line: the longer a battery will last — because it's the right battery for the job and it's well made and well cared for — the better!

Sealed Batteries

Grid-connected systems with battery backup often incorporate another type of lead-acid battery, known as sealed lead-acid batteries or captive electrolyte batteries. Sealed lead-acid batteries are filled with electrolyte at the factory, charged, and then permanently sealed. This makes them easy to handle. They can be shipped without fear of spillage. They won't even leak if the battery casing is cracked, and they can be installed in any orientation — even on their sides. But most important, they never need to be watered.

Two types of sealed batteries are available: absorbed glass mat (AGM) batteries and gel cell batteries. In absorbed glass mat batteries, thin absorbent fiberglass mats are placed between the lead plates. The mat consists of a network of tiny pores that immobilize the battery acid. These tiny pockets also capture hydrogen and oxygen gases given off by the battery when it is charging. Unlike a flooded lead-acid battery, the gases can't escape. Instead, they recombine in the pockets, reforming water. That's why sealed AGM batteries never need watering.

In gel batteries, the sulfuric acid electrolyte is converted to a substance much like hardened Jell-O by the addition of a small amount of silica gel. The gel-like substance fills the spaces between the lead plates.

Sealed batteries are also known as "maintenance-free" batteries because fluid levels never need to be checked and because the batteries never need to be filled with water. They also never need to be (and should not be!) equalized, a process discussed later in this chapter. Eliminating routine maintenance saves a lot of time and energy. It makes sealed batteries a good choice for grid-connected systems with battery backup. In these systems, batteries are rarely used and hence often forgotten and maintained like those in a grid-connected system. Sealed batteries are also ideal for off-grid systems in remote locations where routine maintenance is problematic — for example, in rarely occupied backwoods cabins or cottages.

Sealed batteries offer several additional advantages over flooded lead-acid batteries. They charge faster and release no explosive gases, so there's no need to vent battery rooms or battery boxes where they're stored. In addition, sealed batteries are much more tolerant of low temperatures. They can even handle occasional freezing, although this is never recommended. Sealed batteries discharge more slowly than flooded lead-acid batteries when not in use. (All batteries self-discharge when not in use.)

Unfortunately, sealed batteries are much more expensive, store less electricity, and have a shorter lifespan than flooded lead-acid batteries. They also can't be rejuvenated (equalized) if left in a state of deep discharge for an extended period of time. During such periods, lead sulfate crystals on the plates begin to grow. Over time, small crystals enlarge. Large crystals reduce battery performance by reducing their ability to store electricity. Batteries then take progressively less charge and have less to give back. Over time, entire cells may die, substantially reducing a battery bank's storage capacity.

Large crystals on the plates of flooded lead-acid batteries can be removed by a controlled overcharge, a procedure known as equalization. Although equalization is safe in unsealed flooded lead-acid batteries, it results in pressure buildup inside a sealed battery. Pressure is vented through the pressure release valve on the sealed battery, causing electrolyte loss that could destroy or seriously decrease the storage capacity of the battery.

So, while maintenance-free batteries may seem like a good idea, they are not suitable for many applications.

Wiring a Battery Bank

Batteries are wired by installers to produce a specific voltage and amp-hour storage capacity. Small renewable energy systems — for example, those used to power RVs, boats, and cabins — are typically wired to produce 12-volt electricity. Many of these applications run entirely off 12-volt DC electricity. Systems in off-grid homes and businesses are typically wired to produce 24- or 48-volt DC electricity. The low-voltage DC electricity, however, is converted to AC electricity by an inverter, which also boosts the voltage to the 120- and 240-volts commonly used in homes and businesses.

Sizing a Battery Bank

Properly sizing a battery bank is key to designing a reliable off-grid system. The principal goal when sizing a battery bank is to install a sufficient number of batteries to carry your household or business through periods when the Sun or wind and Sun (in hybrid systems) are not available.

Battery banks are typically sized to meet the need for electricity for three days. Longer reserve periods — five days or more — may be required for some areas. As noted in Chapter 3, backup fossil-fuel generators are often included in off-grid systems. Backup generators can reduce the size of the battery bank and are used to equalize the batteries. For more details on wiring and sizing battery banks for off-grid systems, you may want to check out *Power from the Sun*. Because batteries are expensive, it's a good idea to make your home as efficient as possible. This will reduce the size of your PV system and battery bank.

Battery Maintenance and Safety

Battery care and maintenance are vital to the long-term success of battery-based renewable energy systems. Proper maintenance increases the service life of a battery. Because batteries are expensive, longer service life results in lower operating costs over the long haul. The longer your batteries last, the cheaper your electricity will be.

Keep Them Warm

Batteries like to be kept warm. For optimal function, batteries should be kept at around 75 to 80 degrees. Batteries like to live at about the temperature I find comfortable. In this range, they'll accept and deliver tons more electricity. Guaranteed!

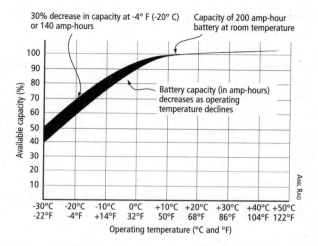

Fig. 7.4: *Battery Performance vs. Temperature. As temperatures drop, battery capacity decreases because cold temperatures dramatically reduce the chemical reactions occurring in batteries. This means you'll get less total electricity from colder batteries.*

Cold temperatures slow down the chemical reactions in batteries and thus dramatically reduce the amount of electricity they store (Figure 7.4).

Although batteries can't be housed in cold rooms, care must be taken to avoid exposure to higher temperatures. High temperatures increase the release of explosive hydrogen gas, a phenomenon known as outgassing. They also increase water loss, which reduces battery fluid levels. Higher temperatures also lead to higher rates of self-discharge in batteries, and the older the battery, the more rapidly it self-discharges.

If you can't maintain batteries in the 75-to-80° F range, at least try to ensure they're housed in a room where the temperature ranges between 50 and 80° F. Rarely should batteries fall below 40° or exceed 100° F. Whatever you do, don't store batteries in a cold garage, barn, or shed. Besides delivering less electricity, they won't last long. They could even freeze under certain conditions, causing their cases to crack. This could lead to a dangerous mess as acid spills out into the room.

Batteries should not be stored on concrete floors. Cold floors cool them down and reduce their capacity. Always raise batteries off the floor.

Ideally, batteries should be housed in a separate, conditioned (heated and cooled) battery room or in a battery box inside a conditioned space to maintain the optimum temperature. Battery boxes are typically built from plywood. An acid-resistant liner is required to contain possible acid spills. Lids should be hinged and sloped to discourage people from storing items on top. As a side note, batteries should be located as close as possible to the inverter and other power conditioning equipment. Doing so minimizes power losses.

Ventilate Your Batteries

Batteries release potentially explosive hydrogen gas when being charged, so battery boxes and battery rooms should be well ventilated (Figure 7.5). This allows hydrogen to escape. Never place

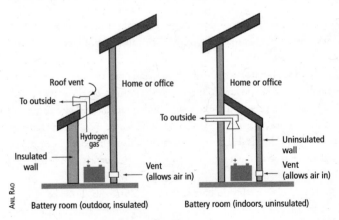

Fig. 7.5: *Battery Vent System. (a) An outdoor battery room should also be well insulated and possibly heated and cooled to maintain temperatures in the optimum range. A passive vent system is needed to allow hydrogen gas to escape. (b) Indoor battery rooms need not be insulated, but they require venting.*

batteries in a room with a gas-burning appliance or an electrical device, such as an inverter, even if the enclosures are vented. A tiny spark could ignite the hydrogen gas, causing an explosion.

Keep Kids Out

Battery rooms and battery boxes should be inaccessible and locked if young children are present. This will prevent children from coming in contact with the batteries, and risking electrical or acid burns. Although electrocution is not a hazard at 12, 24, or 48 volts, dropping a tool or other metal object on the battery terminals can result in an electrical arc that can cause burns, or could result in an explosion of a battery case resulting in acid burns.

Avoiding Deep Discharge to Ensure Longer Battery Life

Keeping flooded lead-acid batteries warm and topped off with distilled water ensures a long life span and optimum long-term output. Longevity can also be ensured by keeping batteries as fully charged as possible. Like many technologies, lead-acid batteries last longer the less you use them. That is to say, the fewer times a battery is deeply discharged, the longer it will last. As illustrated in Figure 7.6, a lead-acid battery that's frequently discharged to 50% will last for slightly more than 600 cycles, if recharged after each deep

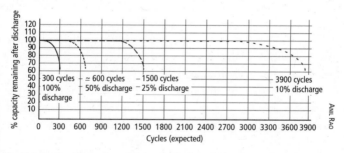

Fig. 7.6: *Battery Life vs. Deep Discharge. Shallow discharging prolongs battery life as explained in the text, while deep discharging reduces battery life. Note the difference in the number of cycles a battery can undergo at routine 50% discharge vs. 10% discharge.*

discharge. If discharged no more than 25% of its rated capacity —
and recharged after each deep discharge — the battery should last
about 1,500 cycles. If the battery is discharged only 10% of its
capacity, it will last for 3,600 cycles.

This topic (like so many others) is complicated. While deep
discharging reduces the lifespan of a battery, renewable energy users
are more concerned with the cost of the battery per watt-hour
cycled. In other words, what we want from batteries is not simply
for them to last a long time but to cycle a lot of energy. So, even
though it goes against the "shallow cycling makes batteries last
longer" idea, I have found that you'll get the most bang for your
buck by cycling in the 40 to 60% deep discharge range.

It's very important to recharge batteries as quickly as possible,
after deep cycling. For long life, you should never leave batteries at
a low state of charge for a long time. This results in the formation
of large lead sulfate crystals, as described earlier. Unfortunately,
achieving these goals is easier said than done. If your system is small
and you don't pay much attention to electrical use, you'll very likely
overshoot the 40 to 60% mark time and time again.

One way of reducing deep discharging is to conserve energy
and use electricity as efficiently as possible. Conserving energy
means not leaving lights and electronic devices on when they're not
in use — all the stuff your parents told you when you were a kid.
It also means ridding your home or business of phantom loads.
Energy efficiency means installing energy-efficient lighting, appliances, electronics, and so on — the actions energy conservation
experts have been suggesting for decades.

Conserving energy and making your home or business energy
efficient is only half the battle, however. You may also have to adjust
electrical use according to the state of charge of your batteries. In
other words, you have to cut back on electrical usage when batteries are more deeply discharged and shift demand for electricity to
times when the batteries are more fully charged. You may, for
instance, run your washing machine and microwave when the
Sun is shining and your batteries are full, but hold off when it's

cloudy and batteries are running low — unless you want to run a backup generator.

To keep track of your batteries' state of charge so you can manage them better, it is wise to install a digital amp-hour or watt-hour meter. These meters keep track of the number of amp-hours of electricity stored in a battery bank each day. (That is, the number of amp-hours produced by the renewable energy system.) It also indicates the number of amp-hours drawn from the batteries. In addition, this meter keeps track of the total amount stored in a battery bank at any one time, that is, how full the batteries are. This information can be used to adjust activities. If batteries are approaching the 40 to 60% discharge mark, you'll want to hold off on activities that consume lots of electricity. Or, you may want to run your backup generator to charge the batteries.

Watering and Cleaning Batteries

To maintain batteries you must also periodically add distilled water. This replaces water lost by electrolysis, the splitting of water molecules in the electrolyte. This occurs when electricity flows into a battery during recharge. Electricity splits water molecules into hydrogen and oxygen gases. (Electrolysis is the source of the potentially explosive mixture of hydrogen and oxygen gas that makes battery room venting necessary.) These gases can escape through the vents in the battery caps on flooded lead-acid batteries, lowering water levels. Water can also evaporate through the vents on a flooded lead-acid battery, and a mist of sulfuric acid can escape through the vents during charging, depleting fluid levels.

All of these sources of water loss add up over time and can run a battery dry. When the plates are exposed to air, they quickly begin to corrode. When this happens, a battery's life is pretty well over.

To prevent batteries from running dry and to ensure optimum performance, you must check battery fluid levels regularly. Many experts recommend checking batteries on a monthly basis. Others recommend checking batteries every two to three months. I have found that a one- to two-month checkup works well in my system.

To check fluid levels, unscrew the caps and peer into each cell when the batteries are not charging. Use a flashlight, if necessary — never a flame from a cigarette lighter! Battery acid should cover the lead plates at all times — at a bare minimum a quarter of an inch above the plates. As a rule, it is best to fill batteries to just below the bottom of the fill well — the opening in the battery casing into which the battery cap is screwed.

When filling a battery be sure to only add distilled or deionized water. Distilling and deionizing are different processes that produce similar quality pure water. Never use tap water. It may contain minerals or chemicals that contaminate the battery fluid, reducing a battery's life span.

Batteries should be fairly well charged before topping them off with distilled water. Don't fill batteries, and then charge them, for example, with a backup generator. Overfilling a battery could result in battery acid bubbling out of the cells when the batteries are charged. If fluid level is extremely low, add a little distilled water before charging them.

Electrolyte loss in overly filled batteries not only reduces battery acid levels, it also deposits acid on the surface of batteries. When it dries, the acid forms a white coating. This not only looks messy, but it can also conduct electricity out of batteries, slowly draining them. Battery acid also corrodes metal — electrical connections, battery terminals, and battery cables.

The tops and sides of batteries need to be cleaned periodically with distilled water and paper towels or a clean rag. When cleaning batteries, be sure to wear gloves, protective eyewear, and a long-sleeved shirt — one you don't care about. If you get acid on your skin, wash it off immediately with soap and water.

When filling batteries, be sure to take off watches, rings, and other jewelry, especially loose-fitting jewelry. Metal jewelry will conduct electricity if it contacts both terminals of a battery. Such an event will leave your jewelry in a puddle of metal — along with some of your flesh. One 6- or 12-volt cell can produce more than 8,000 amps if the positive and negative terminals of a battery are connected. In addition, sparks could ignite hydrogen and oxygen gas in the vicinity, causing an explosion. Shorting out a battery can also crack the case, releasing battery acid.

Also be careful with tools when working on batteries — for example, tightening cable connections. A metal tool that makes a connection between oppositely charged terminals on a battery may be instantaneously welded in place. The tool will become red hot and could also ignite hydrogen gas, causing an explosion. Wrap hand tools used for battery maintenance in electrical tape so that only one inch of metal is exposed on the "working" end; that way it can't make an electrical connection. Or buy insulated tools to prevent this from happening.

You may also have to clean the battery posts of your batteries every year or two. To clean the posts, use a small wire brush, perhaps in conjunction with a spray-on battery cleaner, available at hardware stores. To reduce maintenance, coat battery posts with Vaseline or a battery protector/sealer, available at hardware and auto supply stores. This protects the posts and the nuts that secure the battery cables on the posts.

Equalization

To get the most out of batteries, you need to periodically equalize them. Equalization, mentioned earlier, is a controlled overcharge of batteries.

Why Equalize?

Periodic equalization is performed for three reasons. The first is to drive lead sulfate crystals off the lead plates, preventing the formation of larger crystals. As noted earlier, large crystals reduce battery capacity, and they are difficult to remove. They can also flake off, removing lead from the plates.

Batteries must also be periodically equalized to stir the electrolyte. Sulfuric acid tends to settle near the bottom of the cells in flooded lead-acid batteries. During equalization, hydrogen and oxygen gases released by the breakdown of water (electrolysis) create bubbles.

These bubbles mix the fluid so that the concentration of acid is equalized throughout each cell of each battery, ensuring better function.

Equalization also helps bring all of the cells in a battery bank to the same voltage. That's important because some cells sulfate more than others. As a result, their voltage may be lower. A single low-voltage cell in one battery reduces the voltage of the entire string. In many ways, then, a battery bank is like a camel train. It travels at the speed of the slowest camel.

Although equalization removes lead sulfate from plates, restoring their function, some lead dislodges — or flakes off — during equalization, settling to the bottom of the batteries. As a result, batteries lose lead over time and never regain their full capacity.

How to Equalize

Equalizing batteries is a simple process. In those systems with a gen-set for backup, the owner simply sets the inverter to the equalization mode and then cranks up the generator. The inverter controls the process from that point onward — creating a nice, controlled overcharge that removes sulfate from the lead plates, mixes the electrolyte in the cells, and brings all cells to the same voltage. In wind/PV hybrid systems, the operator can also set the controller to the equalize setting during a storm or period of high wind. The controller takes over from there.

How often batteries should be equalized depends on whom you talk to and how hard you work your batteries. Some installers recommend equalization every three months. If your batteries are frequently deep discharged, however, you may want to equalize more frequently. If batteries are rarely deep discharged, they'll need less frequent equalization. For example, batteries that are rarely discharged below 50% may only need to be equalized every six months.

Rather than second guess your batteries' needs for equalization, it is wise to check the voltage of each battery every month or two. If you notice that the voltage of one or two batteries is substantially lower than others, equalize the battery bank. Checking voltage requires a small digital voltmeter.

Fig. 7.7: *Hydrometer. Hydrometers measure specific gravity. Low specific gravity indicates that the battery needs recharging.*

DAN CHIRAS

Another way to test batteries is to measure the specific gravity of the battery acid using a hydrometer (Figure 7.7). Specific gravity is a measure of the density of a fluid. Density is related to the concentration of battery acid — the higher the concentration, the higher the specific gravity. If significant differences in the specific gravity of the battery acid are detected in different cells of a battery bank, it is time to equalize.

Checking the voltage of the batteries and the specific gravity of the cells may involve more work than you'd like and may not be necessary if you pay attention to weather and battery voltage or adhere to a periodic equalization regime. I rarely equalize my batteries during Colorado's sunny summer months, as my batteries are often full to overflowing with the electricity produced by my solar electric panels. In the winter, however, I equalize every two to three months. When batteries run low, I may run the generator for an hour or two to bring the charge up to prevent deep cycling. This is not an equalization, just an attempt to recharge the batteries more often in cloudy weather.

Small wind energy expert Mick Sagrillo recommends using a wind turbine to equalize the batteries of an off-grid hybrid system.

You'd be amazed at how well this works. As a final note on the topic, be sure only to equalize flooded lead-acid batteries. A sealed battery, either gel cell batteries or absorbed glass matt sealed batteries, cannot be equalized! If you try to, you'll ruin them.

Reducing Battery Maintenance

Battery maintenance should take no more than 30 minutes a month. (It takes about ten minutes to check the cells in a dozen batteries but may take 20 minutes to add distilled water to each cell if battery fluid levels are low.) To reduce maintenance time, you can install sealed batteries, although they're best suited for grid-connected systems with battery backup.

Another way to reduce battery maintenance is to replace factory battery caps with *Hydrocaps*, shown in Figure 7.8. Hydrocaps capture much of the hydrogen and oxygen gases released by batteries when charging under normal operation. The gases are recombined in a small chamber in the cap filled with tiny beads coated with a platinum catalyst. Water formed in this reaction drips back into the batteries, reducing water losses by around 90 percent.

Another option is the *Water Miser cap*. They capture moisture and acid mist escaping from batteries' fluid, reducing water loss by around 30 to 75 percent.

Yet another way to reduce maintenance is to install an automatic or semiautomatic battery filling system, shown in Figure 7.9.

Fig. 7.8: *Hydrocaps. These simple devices help reduce battery watering by reducing water losses.*

JOE SCHWARTZ

I use a manually operated Qwik-Fill battery watering system manufactured by Flow-Rite Controls in Grand Rapids, Michigan, and
sold online through Jan Watercraft Products. This system works
extremely well even after many years of service; it can turn battery
maintenance from a chore to a pleasure.

Although battery-filling systems work well, they're costly. I
spent about $300 for my 12 batteries. Although that's a bit pricey,

Fig. 7.9: *Battery Filling System. Distilled water can be fed automatically to
battery cells or manually pumped into them through plastic tubing. Both
approaches save a lot of time and energy and help to keep battery fluid levels
topped off to ensure battery longevity.*

Fig. 7.10: *Battery Filler
Bottle. If you can access
your batteries relatively
easily, this filler bottle is
one of the easiest and
most economic means of
adding distilled or deionized water to them.*

the systems quickly pay for themselves in reduced maintenance time and ease of operation. The convenience of quick battery watering overcomes the procrastination that leads to costly battery damage. A cheaper alternative is a half-gallon battery filler bottle (Figure 7.10).

Charge Controllers

Now that you understand how batteries work and how to take care of them, let's turn our attention to the charge controller, another device that helps us care for our batteries.

A charge controller is a key component of battery-based PV systems. A charge controller performs several functions, the most important of which is preventing batteries from overcharging (Figure 7.11).

How Does a Charge Controller Prevent Overcharging?

To prevent batteries from overcharging, a controller monitors battery voltage at all times. When the voltage reaches a certain pre-determined level, known as the *voltage regulation* (VR) *set point*, the controller

Fig. 7.11: *Charge controllers like this one (top left) from Apollo Solar regulate the flow of electricity to the batteries in off-grid and grid-connected systems with battery backup. Some charge controllers contain maximum power point tracking circuitry to optimize array output and other features as well, like digital meters that display data on volts, amps, and electricity stored in battery banks.*

APOLLO SOLAR

either slows down or terminates the flow of electricity into the battery bank (the charging current), depending on the design. In some systems, the charge controller sends surplus electricity to a diversion load (Figure 7.12). This is an auxiliary load, that is, a load that's not critical to the function of the home or business. It is often a heating element placed inside a water heater or wall-mounted resistive heater that provides space heat. In PV systems, excess power is often available during the summer months during periods of high insolation. In these instances, the diversion load may consist of an irrigation pump or a fan to help exhaust hot air from a building. Diversion loads must be carefully sized according to the National Electrical Code, something an installer will be sure to do.

Why Is Overcharge Protection So Important?

Overcharge protection is important for flooded lead-acid batteries and sealed batteries. Without a charge controller, the current from a PV array flows into a battery in direct proportion to irradiance, the amount of sunlight striking it. Although there's nothing wrong with that, problems arise when the battery reaches full charge.

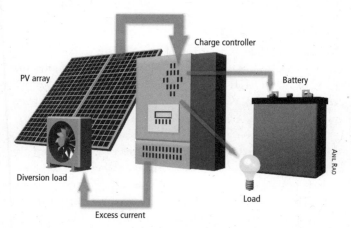

Fig. 7.12: *Diversionary charge controllers send surplus electricity to a dump load, either a resistive heater or fan or pumps, as explained in the text.*

Preventing Reverse Current Flow

At night when the PV array is no longer producing electricity, current can flow from the batteries back through the array. To prevent this reverse current flow, charge controllers contain a diode in the circuit. Were it not prevented, reverse current flow could slowly discharge a battery bank. In most PV systems, battery discharge through the modules is fairly small, and power loss is therefore insignificant. However, reverse current flow is much more significant in larger PV systems. Fortunately, nearly all charge controllers deal with this potential problem automatically.

Without a charge controller, excessive amounts of current could flow into the battery, causing battery voltage to climb to extremely high levels. High voltage over an extended period causes severe out gassing, water loss, and loss of electrolyte that can expose the lead plates to air, damaging them. It can also result in internal heating and can cause the lead plates to corrode. This, in turn, will decrease the cell capacity of the battery and cause it to die prematurely.

Overdischarge Protection

Charge controllers protect batteries from high voltage, but also often incorporate overdischarge protection, that is, circuitry that prevents the batteries from deep discharging. When the weather's cold, overdischarge protection also protects batteries from freezing. This feature is known as a *low-voltage disconnect*.

Charge controllers prevent overdischarge by disconnecting loads — active circuits in a home or business. Overdischarge protection is activated when a battery bank reaches a certain preset voltage or state of charge but only protects against deep discharge caused by DC circuits. This feature prevents the batteries from discharging any further. Overdischarge not only protects batteries, it can protect loads, some of which may not function properly, or may not function at all at lower than normal voltages.

Generators

Another key component of off-grid systems is the generator (Figure 7.13). Generators (also referred to as "gen-sets") are used to charge batteries during periods of low insolation. They are also used to equalize batteries and to provide power when extraordinary loads are used — for example, welders — that would exceed the output of the inverter. Finally, gen-sets may be used to provide backup power if the inverter or some other vital component breaks down. Although a battery-charging gen-set may not be required in hybrid systems with good solar and wind resources, most off-grid homes and businesses have one.

Gen-sets for homes and businesses are usually rather small, around 4,000 to 7,000 watts. Generators smaller than this are generally not adequate for battery charging.

Generators can be powered by gasoline, diesel, propane, or natural gas. By far the most common gen-sets used in off-grid systems

Fig. 7.13: *Portable gen-sets like these commonly run on gasoline.*

are gasoline-powered. They're widely available and inexpensive. Gas-powered generators consist of a small gas engine that drives the generator. Like all generators, they produce AC electricity.

Gas-powered generators operate at 3,600 rpm and, as a result, tend to wear out pretty quickly. Although the lifespan depends on the amount of use, don't expect more than five years from a heavily used gas-powered gen-set. You may find yourself making an occasional costly repair from time to time as well.

Because they operate at such high rpms, gas-powered gen-sets are also rather noisy; however, Honda makes some models that are remarkably quiet (they contain excellent mufflers). If you have neighbors, you'll very likely need to build a sound-muting generator shed to reduce noise levels, even if you do install a quiet model. And don't think about adding an additional muffler to a conventional gas-powered generator. If an engine is not designed for one, adding one could damage it.

If you're looking for a quieter, more efficient generator, you may want to consider one with a natural gas or propane engine. Large-sized units — around 10,000 watts or higher — operate at 1,800 rpm and are quieter than their less expensive gas-powered counterparts. Lower speed translates into longer lifespan and less noise. Natural gas and propane are also cleaner burning fuels than gasoline. Unlike gas-powered generators, natural gas and propane generators require no fuel handling by you, but you could end up paying several times more for a natural gas or propane generator than for a comparable gas-powered unit.

Another efficient and reliable option to consider is a diesel generator. Diesel engines tend to be much more rugged than gas-powered engines and tend to operate without problems and for long periods. Diesel generators are also more efficient than gas-powered generators. Although diesel generators offer many advantages over gas-powered generators, they cost more than their gas-powered cousins. And, of course, you will have to fill the tank from time to time. They're also not as clean burning as natural gas or propane gen-sets.

Living with Batteries and Generators

Batteries work hard for those of us who live off grid. To do their job, they require proper installation, housing, and care. As you have seen, flooded lead-acid batteries need to be kept in a warm place, but not too warm. They need to be installed in vented enclosures and must be kept clean. They also need to be periodically filled with distilled water. And, as if that's not enough, you will need to monitor their state of charge and either charge them periodically with a backup supply of power when they're being overworked or back off on electrical use. Either way, don't allow batteries to sit in a state of deep discharge.

Generators in off-grid systems need a bit of attention, too. If you install one in your system, you will need to periodically change oil and air filters. If you install a manually operated generator, you'll need to fire it up from time to time to raise the charge level on your batteries or to equalize the batteries. It is also a good idea to run a generator from time to time during long periods of inactivity, for example, over the summer when a generator is typically not used. Gasoline goes bad sitting in a gas tank, too, so you may want to add a fuel stabilizer to the tank during such periods. If gasoline evaporates from the carburetor, expect a major repair bill. The residue forms a shellac-like material that really gums up the works. Finally, gasoline-powered generators can be difficult to start on cold winter days, so be sure to use the proper weight oil during the winter.

Gas-powered generators, while inexpensive, tend to require the most maintenance and have the shortest lifetime. Be prepared to haul your gas-powered gen-set in for an occasional repair. My backup generator has been to the repair shop twice in 13 years for costly repairs — and I only runs it 10 to 20 hours a year!

In grid-connected systems with battery backup, you'll have much less to worry about. If you install sealed batteries, for example, you'll never need to check the fluid levels or fill batteries. Automatic controls keep the batteries fully charged.

Batteries may seem complicated and difficult to get along with, but if you understand the rules of the road, you can live peacefully with these gentle giants and get many years of faithful service. Break the rules and it's a sure thing you'll pay for your inattention and carelessness.

CHAPTER 8

MOUNTING A PV ARRAY FOR MAXIMUM OUTPUT

For a solar electric system to operate at its full capacity, it must be installed correctly. The array must be mounted in a location that receives as much sunlight as possible throughout the year. For best results, the array should be oriented so that it points directly to true south in the Northern Hemisphere or true north in the Southern Hemisphere. If the array is fixed, that is, if its tilt angle can't be adjusted, it should be mounted at an angle that provides the maximum output during the year. Or, the array can be installed so that it can be manually or automatically adjusted to optimize output.

Although the task of siting a solar array in a sunny location and orienting the array correctly seems pretty straightforward, it can be quite challenging. For example, it is not always possible to find a location, especially on existing buildings, that provides full access to the Sun all year long. Shade trees, neighboring structures, or even portions of the building on which the array is mounted can shade it part of the time, reducing its output. While every effort should be made to install a PV array in a shade-free location, compromises are often necessary.

Siting an array may also be influenced by aesthetic considerations. Building departments, historic preservation districts, homeowners' associations, and neighbors may influence the decision, putting pressure on you to place an array in a location that could lower its output. This chapter covers proper installation to ensure a safe and productive PV array. It will introduce you to a number of options that provide some flexibility in meeting the goals of maximum output and aesthetics.

What Are Your Options?

Mounting options for the PV array are usually dictated by sunlight/shading issues at the site and by aesthetics. Generally speaking there are two options: roof mounted arrays or ground mounted arrays. Roof or building mounted arrays include "building integrated," where the array is part of the building, and roof-racks. Ground-sited arrays include those that are mounted on top of a steel mast or pole, and those mounted on racks attached to concrete piers secured to the ground. In this chapter, I'll discuss the various options within these two broad categories, beginning with pole-mounted arrays, a type of ground mount.

Pole Mounts — Fixed and Tracking Arrays

A pole-mounted array is a PV array that's attached to a sturdy steel pole anchored securely in the ground (Figure 8.1). The array is typically mounted on the top of the pole, although it can also be mounted on the side of the pole.

Pole-mounted arrays are typically fixed, that is, mounted so the azimuth and tilt angles of the array remain constant throughout

Fig. 8.1:
Fixed Pole-mounted Array. Each array is mounted on a pole anchored to the ground at the optimum angle for its location. Pole-mounted arrays provide some flexibility in placing an array on a lot.

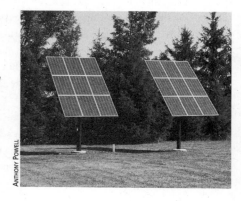

ANTHONY POWELL

the year. Installers choose the optimum year-round angles in such cases. You can obtain data on optimum angles from the NASA Surface Meteorology and Solar Energy website.

To increase the output of your array, you may want to install an adjustable pole-mounted array — also called a seasonally adjustable pole-mounted array. These manually adjusted arrays allow the operator to change the tilt angle of the array by season to increase the array's output. Although you can adjust an array four times a year — spring, summer, fall, and winter — most people make the adjustments twice a year. The first adjustment is made on or around the first day of spring. At that time, the array is adjusted at an angle equal to the site's latitude minus 15 degrees. Reducing the tilt angle positions the array so it captures more of the high-angled summer sun. The second adjustment is made on or around the first day of fall. The array is tilted to an angle equal to the site's latitude plus 15 degrees. This positions the array so that it more directly faces the low-angled winter sun.

Seasonal adjustments of the tilt angle can increase the output of an array by 10% to nearly 40%, depending on the latitude of the site, location of the array on a piece of property, and shading. The adjustment can be as simple as loosening a nut on a bolt on the back of the array mount, then tilting the array up or down (Figure 8.2). A magnetic angle finder is used to get the tilt just right. Once the angle is correct, the nut is tightened.

To see how this works, imagine that you live at 40° north latitude. In the spring, the tilt angle of the array should be set at about 25°. To maximize winter output, the tilt angle of the array should be set at the latitude plus 15 degrees, or, in this example, about 55° in the winter.

Although the rules of thumb work pretty well, I recommend that you check the NASA tables for slightly more precise recommendations. The NASA data is based on actual measurements and factors such as seasonal cloud cover.

A third — and very alluring — option for a pole-mounted array is a tracker, a mechanism that moves the array so that it points

Table 8.1
Monthly Averaged Radiation Incident On An Equator-Pointed Tilted Surface (kWh/m²/day)

Lat 35 Lon -87	Jan	Feb	Mar	Apr	May	Jun
SSE HRZ	2.23	2.93	4.01	4.98	5.52	5.80
K	0.44	0.45	0.48	0.49	0.49	0.50
Diffuse	0.97	1.27	1.66	2.05	2.36	2.47
Direct	3.10	3.45	4.16	4.62	4.76	4.94
Tilt 0	2.20	2.83	3.96	4.95	5.49	5.76
Tilt 20	2.84	3.39	4.42	5.13	5.41	5.56
Tilt 35	3.15	3.61	4.51	4.97	5.05	5.11
Tilt 50	3.30	3.65	4.38	4.58	4.47	4.43
Tilt 90	2.84	2.86	3.04	2.69	2.33	2.21
OPT	3.31	3.65	4.51	5.13	5.52	5.76
OPT ANG	55.0	45.0	34.0	18.0	6.00	2.00

* The entries in the first column represent various tilt angles. The entries in the rows represent peak sun hours each month. The optimum tilt angle for each month is listed in the bottom row.

Fig. 8.2:
The tilt angle of seasonally djustable pole-mounted array can be changed to accommodate the changing altitude angle of the Sun to increase electrical production.

DAN CHIRAS

directly at the Sun (or as close as possible) as it makes its way across the sky. Two types of trackers are available: single-axis and dual-axis.

A single-axis tracker adjusts one angle, the azimuth angle. That is, it rotates the array from east to west, following the Sun from sunrise

Jul	Aug	Sep	Oct	Nov	Dec	Annual Average
5.79	5.28	4.72	3.75	2.60	2.04	4.14
0.51	0.51	0.53	0.53	0.48	0.44	0.49
2.38	2.13	1.71	1.30	1.02	0.88	1.69
5.09	4.85	5.05	4.81	3.71	3.04	4.30
5.76	5.25	4.65	3.72	2.55	2.03	4.10
5.60	5.32	5.08	4.47	3.28	2.70	4.44
5.18	5.08	5.10	4.76	3.63	3.04	4.44
4.53	4.60	4.87	4.80	3.79	3.22	4.22
2.28	2.55	3.15	3.69	3.21	2.84	2.81
5.76	5.35	5.12	4.81	3.79	3.24	4.67
4.00	13.0	29.0	45.0	54.0	58.0	30.1

to sunset. A tracker that adjusts the position of the array to accommodate both the altitude angle and azimuth angle of the Sun is known as a dual-axis tracker. Dual-axis trackers are most useful at higher latitudes because of the long daylight periods in the summer. In the tropics, a single-axis tracker will perform as well as a dual-axis tracker.

Although trackers increase an array's annual output, the greatest impact on energy production occurs during the summer because days are longer and typically sunnier than the rest of the year. Output improves less during the short, often-cloudy days of winter. For those installing grid-tied systems with annual net metering, a tracker's excess summer production helps offset winter's lower production. Trackers are less useful for off-grid systems, however, because summer surpluses are typically wasted. Once the batteries are full, the array's output has nowhere to go.

Trackers track either actively or passively. Active systems use small electric motors to adjust the array angles. In most residential

Fig. 8.3: *This photo sensor sends signals to the controller which adjusts the tilt angle and azimuth angle of the array to track the Sun across the sky.*

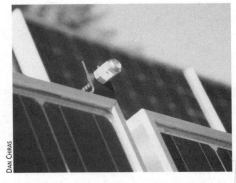

DAN CHIRAS

Fig. 8.4:
(a) This motor sits atop the pole and adjusts the azimuth angle of the array to track the Sun on its east to west path across the sky. (b) A smaller electric motor adjusts the tilt angle of the array to accommodate the ever-changing altitude angle of the Sun.

DAN CHIRAS

a

DAN CHIRAS

b

arrays, these adjustments are controlled by a photo sensor mounted on the array. It sends signals to a small computer (the controller) that activates the motors. Figure 8.3 shows a photo sensor on an array I helped install as part of a workshop. Figure 8.4a shows the electric motor that adjusts the azimuth angle and Figure 8.4b shows the motor that adjusts the tilt angle.

Passive systems require no sensors or motors and are preferred by many installers. They are typically single-axis trackers. As shown in Figure 8.5, passive trackers are equipped with tubes positioned on either side of the array. They are filled with a liquid refrigerant — Freon. When sunlight strikes the tube on the right, it heats the liquid, causing it to expand. Expansion, in turn, forces some liquid into the tube on the left. This causes the weight to shift to the left, which causes the array to rotate, tracking the Sun. Although passive trackers appeal to many who want to minimize moving parts, they do have some downsides. For example, at the end of the day, passive trackers remain pointing west. As the morning Sun rises in the east, it heats the unshaded west-side canister. This forces liquid into the east-side canister. The array moves back to the east. Unfortunately, it takes an hour or two for the passive tracker to return to the east.

An active array, however, returns to its east-facing position right after sunset, so it begins producing electricity an hour or two earlier than a passive array the next morning. As a result, motorized arrays generate more energy over a given period than arrays with passive trackers. Passive trackers can also be deflected by wind, so on windy days they may not be pointing directly at the Sun. They also tend not to work as well in cold climates.

Even though a dual-axis tracker may seem like it would produce a lot more electricity than a single-axis tracker, the benefit is actually only marginal. As shown in Figure 8.5, compared to a fixed array, single-axis trackers generally result in the greatest improvements in array output. Dual-axis trackers increase output over a single axis, but only very slightly.

Active trackers work well, but rely on controllers and other electronic components such as motors and sensors. The more parts, the

Fig. 8.5: *This graph compares the monthly output of a fixed array, a single-axis tracking array, and a dual-axis tracking array. A single-axis tracker dramatically increases the annual output compared to a fixed array. A dual-axis array results in very little additional gain.*

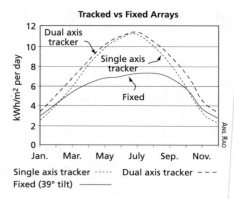

Tracked vs Fixed Arrays

Dual axis tracker

Single axis tracker

Fixed

kWh/m² per day

Jan.　Mar.　May　July　Sep.　Nov.

Single axis tracker ····· 　Dual axis tracker – – –

Fixed (39° tilt) ———

ANIL RAO

more likely that something will go awry. When you add mechanical complexity to an otherwise simple design, you introduce wear, tear, maintenance, and repairs. The more complex the system, the more problems you'll have.

The weakest link in active tracking arrays is the controller. They contain electronic circuitry that is vulnerable to nearby lightning strikes. While manufacturers have improved lightning protection in active trackers, a nearby or direct lightning strike can damage a controller, necessitating replacement.

Although it is true that trackers increase the output of a PV array, you may be better off investing in more PV modules to boost the output of a PV array than installing a tracking array that yields the same output. Both options, may cost about the same. But a fixed array will be practically maintenance free.

One of the biggest advantages of pole-mounted arrays is that they can be positioned far away from objects that might shade an array — just so long as they are not too far from the balance of system. According to PV expert Joe Schwartz, if the modules are wired to 48 volts nominal or higher, an array can be located a couple hundred feet from the inverter or batteries with minimal transmission loss and relatively small gauge wire.

Arrays mounted on poles can also be positioned precisely — that is, pointed south and tilted at just the right angle. Pole-mounts also help to maintain a cooler array than a roof mount can. Most PV modules produce more electricity at cooler temperatures.

Pole-mounted arrays are also a lot safer to install than roof arrays. That's because there's no need for potentially dangerous roof work. Pole-mounted arrays may be your only choice if a roof is unsuitable for some reason — for example, if the roof faces the wrong direction, is at an inappropriate pitch, is shaded, or is not strong enough to withstand wind loading (the force of the wind that could rip an array from a roof). In addition, pole-mounted arrays do not require roof penetrations. Pole mounting an array also avoids having to dismantle an array when it comes time to re-roof

Fig. 8.6 a and b:
Pole-mounted array under construction. (a) Worker attaches the aluminum braces that will support the PV modules (b) to a steel strongback.

a

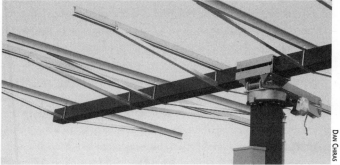

b

Fig. 8.7:

Workers carefully slide PV modules into place on this dual-axis tracker at a hands-on workshop in Minnesota taught by Chris LaForge.

DAN CHIRAS

a house. Generally speaking, PV modules will long outlive most roofs, sometimes multiple re-roofings.

On the downside, pole-mounted arrays are not usually suitable for small lots in cities and suburbs because trees and homes can block the Sun. Rural lots are typically better suited to pole-mounted arrays.

Pole-mounted arrays are vulnerable to strong winds. An array becomes a large sail on windy days. Proper design and installation are critical to prevent wind damage.

Rack Mounts

Pole mounts are growing in popularity, but most PV systems are still mounted on racks. Racks, like the one shown in Figure 8.8, can be mounted on sloped or flat roofs or on the ground. Ground-mounted racks must be secured to a solid foundation. Racks can also be mounted on the south sides of buildings. There, the PV modules serve two functions: they generate electricity, and they shade windows and walls from the intense summer sun, helping to passively cool homes and businesses.

Once made primarily from steel, racks are now manufactured from aluminum. These solid, durable racks secure the array to the roof, wall, or ground, preventing the array from being carried away by winds. They may also provide a means of adjusting the tilt angle to optimize energy production.

Racks consist of three main components: (1) anchors or feet that attach the rack to a solid surface — a rooftop, a wall, or a foundation, in the case of a ground-mounted rack; (2) legs that allow the installer to set and, in the case of adjustable racks, to change the tilt angle seasonally; and (3) horizontal bars called rails to which the PV modules are attached (Figure 8.8).

Racks for solar arrays come in two basic varieties: fixed and adjustable. Fixed racks are typically set at the optimum angle, typically the latitude of a location. If you live at 35° north latitude, for instance, the array should be set so the tilt angle is 35°. Be sure to check with the NASA website for the optimum tilt angle. It may be plus or minus 5 degrees.

Adjustable racks can be adjusted to maximize the annual energy production of an array. As noted earlier, most arrays are adjusted twice a year. Adjustability may be provided by telescoping back legs that allow the rack to be tilted up or down. Some tilting legs come with predrilled holes or slots that correspond to different tilt angles.

Fig. 8.8: *This aluminum rack made by UniRac can be mounted on roofs or on the ground. They may be fixed or seasonally adjusted. Note the mounting clips on the bottom rail.*

Racks are relatively easy to install. They also permit air to circulate around an array, keeping it cooler and increasing its output. On the negative side, racks are subject to the powerful forces of the wind. Rooftop, awning, and, to a lesser extent, ground-mounted racks can be ripped from their moorings by strong winds — if not properly installed.

Roof-mounted racks also require numerous roof penetrations caused by driving lag screws into the roof for rack installation and sometimes drilling holes to run wires to the balance of system (Figure 8.9). If these penetrations are not sealed properly, they can leak, damaging the home and reducing the effectiveness of ceiling insulation. Roof-mounted racks can also be difficult and costly to install in some instances — for example, on homes with steep or extremely high roofs. Expect to pay more if you have one of these difficult-to-access and dangerous roofs.

Roof-mounted PV systems are also more difficult to access than ground-mounted or pole-mounted PV arrays to adjust the tilt angle, brush snow off the modules, or remove dust.

One of the most significant disadvantages of roof-mounted racks is that they must be disassembled and removed when time comes to re-roof a house. This is costly and time-consuming work that should be carried out by a trained professional, not your local roofing contractor.

Removing an array to re-roof a house is also a huge annoyance. Asphalt shingles, found on the majority of homes in the United States live relatively short lives. This is especially true in areas that experience damaging hailstorms, intense sunlight, or extreme temperature changes. So, when contemplating a roof-mounted rack, carefully evaluate the condition of your roof or hire a trustworthy roofer to give you an honest opinion. If your roof needs replacement, or will need replacement soon, do it *before* installing your PV system.

When building a new home or re-roofing an existing home, install a highly durable roofing material. A metal roof that should last for 50 years or more is an excellent choice, if it fits with the

Fig. 8.9: *These L-footings are bolted into a 2 x 4 block attached to adjacent roof rafters. Silicon caulk is used to seal the hole and seal the base of the footing to prevent water from leaking into the roof.*

architectural style of your home. Many of the environmentally friendly roofs made from recycled plastic and rubber also outlast standard asphalt shingle roofs by decades. At the very least, chose high-quality (long-lasting) asphalt shingles. They're more expensive, but could last 40 years.

Standoff Mount

Your third option for mounting an array is known as a *standoff mount*. Shown in Figure 8.10, this special form of roof mount allows installers to mount PV arrays parallel to the roof surface to reduce their profile and visibility. Because the array is mounted parallel to the roof surface, the PV modules are mounted at a fixed tilt angle corresponding to the angle of the roof.

In standoff mounts, the array is raised slightly off the roof by four to six inches, sometimes more. The space between the roof and the array helps keep the PV modules cooler than a direct mount or an array attached directly to the roof (an approach that is rarely, if ever, seen except when laminate PVs are attached to standing seam metal roof).

"Standoff-mounted arrays are the most common, preferred, and least-expensive method for installing arrays as a retrofit or to existing rooftops," according to Jim Dunlop, author of *Photovoltaic Systems*, an excellent textbook on solar electricity for aspiring professionals. The arrays are mounted on rails typically attached to metal feet lagged into the roof framing members (Figure 8.9).

ROCHESTER SOLAR TECHNOLOGIES

Fig. 8.10: *These modules are mounted "flush" to the roof to reduce their profile. They're actually mounted several inches off the roof on a metal rack. This is known as a standoff mount.*

Standoff mounts are popular among homeowners and businesses primarily because they minimize the profile of an array, making it appear as if it is part of the building. This option is not only aesthetically appealing, compared to pole or rooftop racked arrays, but it also reduces the wind's effect on arrays. It's far less likely that an extremely strong wind will dislodge a standoff-mounted array from its anchorage than a conventionally mounted array because the wind passes over the top of the array.

While aesthetically appealing, standoff-mounted arrays have their downsides. One of them is that the pitch (slope) of the roof determines the tilt angle. If the pitch of the roof does not correspond with the optimum tilt angle, the PV array's production could be seriously compromised. If the pitch is too shallow, for instance, wintertime production could be substantially lower than a homeowner could achieve by installing a properly angled rack or pole array. For off-grid systems, reduced output, especially in the winter, can be devastating. To compensate, a homeowner would need to install a larger — potentially much larger — and more costly PV system. If, on the other hand, the pitch is too steep, the array's output in the summer could be compromised.

Another disadvantage of standoff mounts is that they frequently result in higher array temperatures than rack-mounted or pole-mounted arrays. This reduces the array's annual output.

Building-Integrated PVs

In Chapter 3, I discussed the many options for integrating PVs into buildings — for example, in roofs, walls, and glass. For most homeowners and businesses, there are three practical options: solar tiles, PV laminates on standing-seam metal roofs, and solarscapes. Let's take a look at each one.

Laminate PVs for Standing-Seam Metal Roof

If the roof of your home or business is fitted with standing-seam metal roof, or if you're about to re-roof and would like to install this product, another option for turning your roof into an electric-generating plant is to install PV laminate, or PVL (Figure 8.11). Developed by United Solar Ovonic and the National Renewable Energy Laboratory, PV laminate is fabricated by applying multiple layers of amorphous silicon to a durable, but flexible metal backing. It is adhered to flat

Fig. 8.11: *PV laminate is applied directly to a standing seam metal roof as shown here.*

sections of standing-seam metal roofs, between the ridges (standing seams), and can be applied to a new or existing roof — although it is extremely difficult to install it on the roof. It's best to install PVL on the metal sections before they are applied to the building.

Uni-Solar's PV laminate (PVL) comes in rolls. They're installed one at a time on sections of metal roofing. The ridge end of the roll is placed near the ridgeline. The paper backing is then removed from the first 16 inches of the roll, exposing the adhesive that secures the PVL to the roof. This section is then secured to the substrate, being careful to avoid creasing. The rest of the length is then rolled up toward the ridge. The backing is then removed a little at a time, as the PVL is unrolled down from the ridge to the eave.

When applying PVL, care should be taken to keep the laminate as straight as possible as it is unrolled. Uni-Solar's adhesive is pretty strong. You can't lift and reposition PVL if you mess up.

In new construction, PVL should be applied to the standing seam metal roof *before* the metal roofing is installed. This can be performed indoors on a temporary workbench. Doing so gives you more control over the application and could result in a better installation. Be sure to follow the manufacturer's instructions.

Once the laminate is installed, the wire leads from the roof panels must be connected. Leads can then be run under the ridge cap if there is an attic or under eaves in buildings with vaulted ceilings.

PVL blends with the building and from a distance may not visible. Although PVL is flush mounted, thin-film products are less vulnerable to high temperatures than crystalline PVs — so, their output doesn't decrease as much at high temperatures as monocrystalline and polycrystalline PVs. They're also not quite as sensitive to shading as standard PV modules. PVLs tend to be cost competitive with rack-mounted arrays.

On the downside: because PVLs are less efficient than crystalline PVs, you'll need twice as much roof space for the same system. Standing seam metal roof is, in our view, also not as attractive as other roofing materials. It's rarely used on homes, which limits the application of this product.

Solar Tiles

Individuals interested in BIPVs (building integrated photovoltaics) may also want to consider solar tiles, like those manufactured by Sharp, Atlantis Energy Systems, and Kyocera. Solar tiles incorporate crystalline silicon cells in a tile that's attached directly to the roof of homes and offices (Figure 8.12). In SUNSLATES™, the single-crystal PV cells are mounted on a fiber cement slate backing. Kyocera's and Sharp's solar tiles are made from polycrystalline PV cells.

Like solar shingles, solar tiles do not need to cover the entire roof. You can, for instance, dedicate a portion of a south-facing roof to your array. The size of the system depends on the solar resource, roof pitch, shading, and electrical demand. As in all solar electric systems, the ideal location is a south-facing roof; however, solar tiles can also be mounted on roofs that face southeast or southwest with only a modest loss (2 to 4%) in production, according to some manufacturers. Mounted on east- and west-facing roofs, the output could decline by 10 to 15%.

Solar tiles shed water from the roof but cannot be mounted on flat roofs. Roof pitch for SUNSLATES, for instance, must be at

Fig. 8.12: *SUNSLATES from Atlantis Energy Systems are solar tiles that replace ordinary roof shingles. The top of this roof is fitted with solar tiles that generate electricity to help meet household demands.*

least 18 degrees (4/12 pitch). Be sure to check the pitch of your roof and manufacturers' recommendations before committing to a solar roof tile.

Solar tiles work well in a variety of climates, but perform best in sunnier locations. Be sure to check with the manufacturer for advice on the suitability of their product for your location. Also, be sure to check the compatibility of a solar tile with your roof.

Solar roof tiles are mounted on roofs over 30-pound roof felt. Each manufacturer provides directions. Solar tiles offer many of the advantages of solar roofing products. The main benefit is aesthetics. They also have some of the same downsides, the most important of which is that they require a lot of connections. Many connections not only mean more work, it also means that these arrays are more difficult to troubleshoot if something goes wrong.

Solarscapes

A third and increasingly popular option designed to integrate solar electricity into new or existing homes and businesses is referred to as *solarscaping*. In this approach, PV modules are incorporated into the roofs of structures, such as gazebos and carports. PV modules can be used to create awnings or shade structures for decks and hot tubs. The PV modules generate electricity while creating shade (Figure 8.13).

Solar shade structures are typically made from wood or steel and are designed and built so they blend in nicely with the existing architecture and landscape. These custom-made structures are erected in sunny locations where shade is desired or required.

Many installers use bifacial PV modules like those made by Sanyo. Bifacial modules were discussed in Chapter 3. This module was developed in the mid 1970s for use in spacecraft. Bifacial modules contain PV cells encapsulated — both front and back — by glass. This allows the module to harvest solar energy from both sides. Sunlight reflecting either off the ground or off light-colored materials, such as light-colored shingles, concrete, gravel, snow, and water, strike the backside of the PV cells, generating additional electricity.

Fig. 8.13: *PV modules can be used to create roofs of shade structures, a technique known as solarscaping. This photograph shows a portion of a carport made from PV modules by Lumos Solar.*

The glass-on-glass construction of bifacial modules allows some light to filter through the array, creating "a soft light-and-shadow pattern on the surfaces beneath the array," according to writer, Topher Donahue in an article on solarscaping in *Home Power* (Issue 122).

Solarscaping is suited for new and existing homes and businesses. Like other types of BIPV, this approach helps minimize aesthetic concerns of conventional PV mounting systems, specifically racks and pole mounts. It therefore helps more people find a suitable way to incorporate PVs into their lives.

Solarscaping also offers the flexibility of pole-mounted and ground-mounted arrays — that is, it allows placement of arrays in sunny locations. In addition, this approach offers ease of access to the array and ease of installation compared to roof-mounted arrays. And, by avoiding installation on the roof, you won't have to worry about removing the array when time comes to replace shingles. Like pole-mounted arrays, this approach results in a cooler and more energy-efficient array. Homeowners and businesses can even buy prefabricated solarscape kits from the Colorado-based company Lumos Solar (www.lumossolar.com).

Conclusion

You have many options when it comes to mounting a PV array. When considering your options, remember that one of your main

Getting the Most from Your PV System

1. Optimize electrical output by locating your PV array in the sunniest (shade-free) location on your property, orienting the array to true south, setting modules at the optimum tilt angle or installing them on an adjustable rack or tracker. Ensure access to the Sun from at least 9 am to 3 pm.

2. Install a Maximum Power Point Tracking controller (for battery-based systems) or inverter (for grid-connected systems) to ensure maximum array output.

3. Keep modules as cool as possible by mounting them on a rack or pole, ensuring air circulation around the modules. Avoid flush-mounts (standoffs), if possible. In hot climates, install high-temperature modules, that is, modules that perform better under higher temperatures.

4. Install high-efficiency modules if roof or rack space is limited.

5. Install modules with bypass diodes to reduce losses due to inadvertent shading.

6. When installing multiple rows of arrays, be sure to provide a sufficient amount of distance between the rows to avoid shading — that is, one row shading the row behind it.

7. Select modules with the lowest rated power tolerance. Power tolerance is expressed as a percentage, which indicates the percent by which a PV module will overperform (produce more power) or underperform (produce less power) the nameplate rating. Look for ☞

goals is to produce as much electricity as possible. The accompanying sidebar, "Getting the Most from Your PV System," summarizes the many ways you can achieve this goal. Optimizing a PV array not only optimizes the value of your investment, it helps create a more sustainable supply of energy. But don't forget aesthetics and curb appeal. Although you may like the looks of a PV

modules with a small negative or a positive-only power tolerance.

8. Install an efficient inverter. Use the weighted efficiency as your guide rather than maximum efficiency. Weighted efficiency is a measure of the efficiency under typical operating conditions.

9. Keep your inverter cool. If installed outside, be sure the inverter is shaded at all times. If installed inside, put it in a cool location. Be sure that air can circulate around the inverter.

10. Decrease line losses by installing a high-voltage array. Reduce the length of wire runs whenever possible. Use larger conductors to reduce resistance losses. Use quality (low-resistance) connectors and equipment. Be sure all connections are tight to reduce resistance and conduction losses.

11. Keep your modules clean and snow free. Brush snow off gently — never use any sharp tools to remove snow from a PV module. If you live in a dusty environment with very little rain, periodically dust off your modules. Mount the array so it is easily accessible for cleaning and/or snow removal.

12. For battery-based systems, maintain batteries properly. Periodically equalize flooded lead-acid batteries and fill them with distilled water. Recharge promptly after deep discharges. Install batteries in a warm location so they stay at 75 to 80°F, if possible. ■

array mounted on a rack on a roof, neighbors might not share your views. Future buyers may not like the look of it either.

Also, keep in mind the ease of installation. The more difficult and risky the installation, the more costly it will be. Although PV arrays require very little maintenance, access is especially important if you live in a dusty or snowy area. Arrays may need to be dusted off or snow may need to be removed occasionally. Also, don't forget about winds and protecting an array from vandals and thieves.

Finally, solar electric arrays must be properly installed and must comply with all requirements of the local building department. The electrical portion of the installation must comply with the National Electrical Code and any local codes. If possible, you may want to consider making your array perform double duty, generating electricity and providing shade.

CHAPTER 9

FINAL CONSIDERATIONS:
PERMITS, COVENANTS, UTILITY
INTERCONNECTION, AND BUYING A SYSTEM

To those who are drawn to the idea of generating their own electricity from the Sun, there are few things in the world more exciting than turning a PV system on for the first time and watching the meter show that electricity generated by the Sun is flowing into your home or business. It's an even greater thrill to see the utility meter running backward as surplus flows onto the electrical grid.

Although you may encounter a few obstacles along the way, in most instances, the path from conception to completed installation is fairly straightforward. Table 9.1 summarizes the steps you or your installer must take — in the order in which they must be taken. As you can see, I've already discussed steps 1 and 2 and details of others. In this chapter, I'll explore the remaining steps. We will begin by discussing permits, then look at restrictive covenants imposed by some homeowners' and neighborhood associations. We will then discuss interconnection agreements required for grid-connected PV systems and how to insure your system. We'll end with some advice on buying and installing a PV system.

Permitting a PV System

After you or your installer have designed your PV system, you'll need to contact your local building department to determine if they require a permit to install a PV system. In most cases, a permit will

Table 9.1
Steps to Implement a PV Energy Project

1. Determine your home or business' electrical consumption and consider making efficiency improvements to reduce the PV system size and cost.
2. Assess the solar resource.
3. Size and design the system.
4. Check on building permit requirements; file a permit application.
5. Check homeowner association regulations; file necessary paperwork for permission to install the system, if required.
6. Apply for special incentives that may be available from the utility or your state or local government.
7. Check on insurance coverage.
8. Contact the local utility and obtain utility interconnection agreement (for grid-connected systems).
9. Obtain permit.
10. Order modules, rack and balance of system.
11. Install system.
12. Commission — require installer to verify performance of the system.
13. Sign interconnection agreement.

be required. The local building department will provide a permit application.

Building departments are typically the "authority having jurisdiction" (AHJ) over all aspects of construction in cities, towns, and counties. They are granted authority and legal power to administer, interpret, and enforce building codes by local governments. Building codes are a detailed set of regulations that apply to how buildings are constructed, modified, and repaired. Building codes set the rules by which the various trades must operate. They stipulate equipment and materials that can be used and how they must be installed. For example, they stipulate the maximum voltage at which PV systems can be wired and the use of safety measures such as disconnects.

Local trade licensing codes (separate from building codes) may also stipulate who can perform certain tasks, for example, licensed electricians or plumbers. Building codes also stipulate setbacks — which limit how close a PV system may be installed to a public right-of-way or neighbors' property lines.

Building departments issue permits for structural and electrical work separately and conduct inspections at certain stages of the work. When a project is completed and passes all required inspections, the local AHJ grants a certificate of approval for PV installations.

Rather than create their own building codes, most AHJs have adopted model building codes, created by independent organizations. In the United States, for instance, most jurisdictions have adopted the National Electrical Code (NEC) for all electrical work on homes and businesses. The National Electrical Code was developed by the National Fire Protection Association's Committee on the National Electrical Code. Article 690 of the NEC applies specifically to the installation of small scale photovoltaic systems.

Although states and cities typically adopt a model code, they are given the authority to modify the model code to meet local or regional needs. AHJs may waive certain requirements or permit alternative measures or equipment that ensures the same safety standards. Even so, AHJs rely almost entirely on the stipulations set forth in the NEC.

Building code requirements not only dictate how buildings are built, they also influence the manufacturers of electrical equipment, including PV systems. Manufacturers of all components of PV systems — from PV modules to inverters and DC disconnects — comply with code requirements — so the equipment they produce meets standards of the National Electric Code. (Equipment that doesn't meet the standards can't legally be installed in areas where building codes are enforced.) To meet NEC requirements, equipment must be approved by an independent testing laboratory, the largest of which is Underwriters Laboratories (UL). It tests equipment and, if the equipment passes muster, then lists or certifies the product as meeting a certain standard. The UL listing is included on the nameplate of the product. UL-listed products may comply with US or Canadian safety standards, or both.

AHJs require the use of approved equipment in PV systems; plan examiners review permit applications, and inspectors verify that the proper equipment was used and that the work was done in compliance with the electrical code. Inspectors are licensed and certified by either the state or local government and have a thorough knowledge of building codes and construction, in the case of building inspectors, or electrical wiring, in the case of electrical inspectors.

Building codes ensure that all building projects, including the installation of PV systems, are safe for the immediate occupants of a home or business and as well as for future residents of a home or employees of a business. The NEC, for example, protects against shocks and fires that could be caused by electricity. If you install a PV system not in accordance with the local electrical code and it causes a fire that destroys your house, your insurance company may deny your claim. The NEC provisions that apply to PV systems also help ensure the safety of utility company employees, notably, line workers, in the case of utility-connected systems. In a sense, then, building codes are society's way of ensuring the safety and well being of the present and future generations.

Although the requirements of local building codes can be daunting to the uninitiated, professional installers should be intimately familiar with them.

Securing a Permit

To start the process of ensuring that your PV installation will comply with local building code, you or your installer must submit an application for a permit. Permits for PV systems encompass electrical wiring and all the electronic components of a system you'll be installing, such as inverters, charge controllers, and safety disconnects. Your application may also be required to describe how the mounting of the array meets appropriate codes — for example, that an existing roof can support a roof-mounted array and that it is securely attached to the roof so it won't blow away in the wind.

To determine if a permit is required, even for off-grid systems, give your local building department a call. If a permit is required, they'll outline the procedure, indicate the cost, and provide the appropriate forms. They will also indicate all the supporting material you'll need to provide as part of your application.

Applications generally require a site map, drawn to scale, that indicates property lines, streets, and the proposed location of the array. This simple drawing should also indicate the location of other components of the system, for example, the inverter, disconnects, and electric meter (in grid-connected systems). A professional installer will prepare a site map for you. It's part of his or her service.

Building permit applications for PV systems usually require a simple one-line drawing of the electrical system, or, less commonly, a more complex three-line electrical diagram. Electrical schematics such as these indicate all of the components of PV systems and the connection with a home's or business' electrical system. You'll also need to submit specifications — size and ratings — of all the system components, either on the system diagram or on separate specification sheets. This includes wire size, overcurrent protection devices (fuses or circuit breakers), system disconnects, and grounding equipment. You'll also need to include specifications for PV modules and inverters in grid-connected systems and charge controllers and batteries in battery-based systems. The building department will use this information to determine if the conductors, overcurrent protection devices, and disconnects are sized correctly.

Even if your local building department does not require an electrical permit, which is rare, the local utility may require an electrical drawing and specifications if you are connecting to the grid. They will examine the diagrams and specification sheets to be sure they comply with the NEC.

You will also very likely need to provide a description and/or drawings of the array mounting design when applying for a permit through the local building department. If you are mounting an array on a roof, you may need to include information on the age of the roof, type of shingles, the pitch, and the size and spacing of the rafters. You'll need to provide information on the array, too, including the weight and the method of securing it to the roof. You'll need to provide details on waterproofing the attachment as well. You may need to hire a professional engineer to review and stamp your plans. An engineer's stamp ensures that the roof can support the array and that it will remain intact under wind conditions in your area. If you're mounting an array on a pole mount, you will very likely need to provide a description and drawings of the foundation and pole mount to ensure its structural stability.

If you are hiring a professional installer, he or she will file the permit and take care of these details. If you're installing a system yourself, however, you'll have to be sure that your plans comply with the local code and provide all the information required by the building department. If that sounds like too much, you may want to consider buying a packaged kit, which includes all of the components you'll need, including diagrams of the system and specifications of the components. Kits make it easier on the do-it-yourselfer.

After your application is submitted, a plan examiner in the building department will review the application and accompanying materials, usually within a few days or weeks, although the process can take several months, depending on the workload and staffing of the building department. If everything is in order, you'll receive a permit, which is the official approval for you to begin construction. (Never order equipment or start work until you've received your permit!)

If your permit application doesn't meet the building code in some way, the plan inspector will mark up the schematics — it's called *redlining* — or provide written notice indicating the problem or problems and required changes. Once the problems have been resolved, you can resubmit the permit application.

Fees for permits for residential PV systems usually run from $50 to $1,500, depending on the jurisdiction. Some jurisdictions charge a flat fee; others charge a fee based on the size of the system.

Permits must be posted on the site, usually in a window. Permits include an inspection schedule. That indicates which parts of the project are subject to inspection, what is covered in each inspection, and the sequence in which inspections must be carried out.

Expect an electrical inspector to visit your site at least once to check wiring and warning labels. (Warning labels are required on disconnects and other components of PV systems). The inspector will check out wire sizes, connections, overcurrent protection, and grounding.

All equipment must be readily accessible to inspectors and must be installed according to code with proper clearances to ensure room to work on the equipment. This is known as *working space*. You'll also need to be sure that no other wiring, plumbing, or ductwork is within a certain distance from your installed equipment.

A structural inspector may also visit to check out the array mount. You or the installer must call the AHJ to schedule all inspections, usually 24 to 48 hours in advance.

If you fail an inspection, you'll need to fix the problem and arrange a follow-up inspection. You may have to pay for follow-up inspections, too, although the cost may be included in the initial permit fee.

While I am on the subject, avoid the tendency to install a PV system without necessary permits. The consequences are too great. Municipal governments have the authority to force homeowners to remove expensive unpermitted systems. Be sure to secure an interconnection agreement with the local utility, too. Without it, you can't connect.

Permitting a PV system may take several months, so submit your application well in advance of the date you'd like to install the system, *and don't buy a PV system until your permit has been granted*. And, as a final note on the subject, keep a copy of the certificate of approval. You may need it in the future if you file an insurance claim.

Covenants and Neighborhood Concerns

Many subdivisions are governed, in part, by restrictive regulations, usually referred to as *covenants*. Covenants are agreements incorporated into the document under which subdivisions are created.

They require or prohibit certain activities. These covenants "run with the land," meaning that they apply to the original owner and all subsequent owners of the property. Some covenants give the neighborhood association the right to create and enforce additional rules and architectural standards.

Restrictive covenants and neighborhood association rules address many aspects of our lives — from the color of paint we can use on our homes to the installation of a privacy fence. Some expressly prohibit renewable energy systems, such as solar hot water systems and solar electric systems.

Restrictive covenants are legally enforceable and the courts have consistently upheld their legality. To find out if you will be prohibited from installing a PV system, review both the restrictive covenants and neighborhood association rules, if they exist.

If your subdivision has restrictive covenants or architectural standards, you need to apply for permission to install a PV system. The application is often a written letter with a drawing or two. Precedence can help. In other words, if someone else has installed a similar system, even without permission, it's easier to obtain approval.

Covenants can hamper renewable energy, so some renewable energy-friendly states like Colorado and Wisconsin prohibit homeowners associations from preventing homeowners from installing solar and other renewable energy systems. If you're in one of those states, count yourself lucky. If not, you might want to work to pass such legislation in your state.

Connecting to the Grid: Working with Your Local Utility

When considering installing a grid-connected system, be sure to contact your local utility company early on — *before* you purchase your equipment — to work out an arrangement to connect to their system. You'll need to file an application that provides sufficient detail to assure the utility that the electricity your PV system will be backfeeding onto the grid will be of the same quality as grid power. They will also want assurance that your system won't continue to backfeed electricity onto the grid if the grid goes down (to protect workers and neighbors). Their contract will stipulate their policy on compensation for any monthly or annual surplus electricity (net excess generation).

Fortunately, many states (43 at this writing) have enacted net metering policies that credit customers for the electricity they feed onto the grid at the same rate the utility charges (retail rates) up to the point of surplus. As noted in Chapter 5, state laws vary with respect to their treatment of monthly and annual net excess generation. In some states, the surplus is granted to the utility. In others, the utility pays wholesale (avoided cost) for the surplus; in a few, the utility pays the customer at their retail rate.

Some utilities use existing electrical meters to keep track of the ebb and flow of electricity, provided they are capable of spinning backward and forward. (As noted in Chapter 5, these meters spin backward when electricity is being backfed onto the grid and forward when electricity is flowing from the grid to your home or business.) If your meter is not bidirectional , the utility may require installation of a second meter to track the electricity you sell to them. Or, they may require a digital meter that keeps track of the flow of electricity to and from the grid. If a new meter is required, you'll often be expected to pay for it.

Some utilities also charge an interconnection fee. This fee may be based on the size of the system or may be fixed — with one price covering all systems. Fees typically range from $20 to $800. They cover costs borne by the utility, such as the cost of inspecting a PV system after it is installed. Be sure to ask about interconnection fees upfront and be sure the contract is clear about any additional fees that may be charged to you over the course of the contract. Also, be sure to ask whether the utility is claiming renewable energy certificates (RECs) for your PV system. They usually do. RECs are financial instruments that can be sold to individuals wishing to offset their carbon emissions. They can be a valuable financial asset to some customers, especially those businesses that install PV arrays. They can be sold to a number of third parties involved in brokering green energy certificates.

When installing a grid-connected system, most installers contact the utility at the same time they file for a building permit. Unless your local utility has lots of experience with solar electric systems, they probably won't have a specific person who deals with interconnection agreements. If your utility is one of the more experienced companies, however, they may have a point person who deals with such matters. Check out your utility's website and look for terms such as "net metering," "interconnection agreement," "renewable energy systems," or "distributed generation." You may find all the information you need online.

Although utilities are required to allow small electrical generators to connect to the grid, getting them to cooperate may be another matter entirely. Problems are especially common in rural electric cooperatives, many of which have been hostile to the idea of allowing small-scale renewable energy generators to connect. So, unless you know that your utility supports customer-owned renewable energy projects, call your state's public utility commission (PUC) before calling your utility. Their staff should be able to describe the approval process for the various types of utilities (publicly owned, rural electric, and municipally operated). They may also put you in touch with the appropriate contact person at the utility. When working with the utility, be prepared to answer all of that person's questions and to fill out the application fully.

Before they sign a contract with you, the utility will very likely require you to submit a site plan indicating where components are located and electrical schematics. Its two most important concerns, however, will be an assurance that you have liability insurance and that your system includes an automatic disconnect. All utility-connected inverters on the market incorporate an automatic disconnect feature. Even so, the utility may require an inspection of the inverter and a demonstration.

As noted in Chapter 5, utilities may also require installation of a visible, lockable disconnect — that is, a manually operated switch that allows utility workers to disconnect your system from the outside of your home in case the grid goes down and they need to work on the electrical lines. (Note that utilities may reserve the right to disconnect your PV system to work on their system without notifying you.) Because grid-connected inverters automatically disconnect from the grid when they sense a drop in line voltage or a change in frequency, this requirement is unnecessary. Regardless, it may still be required by your local utility and will add to your installation costs.

Making the Connection

Once a grid-connected PV system has passed final inspections by the local AHJ, you or the AHJ must contact the utility. At that point, they will sign the interconnection agreement, granting you the right to connect to the grid. You can flip the switch and start generating electricity. In some cases, however, utilities insist on inspecting and testing PV systems themselves to be sure the automatic disconnect functions as it is supposed to.

Insuring Your System

PV systems are expensive and should be insured. Two types of insurance are required: property damage which protects against damage *to* the PV system; and liability insurance, which protect you against potential damage *caused by* the system. Both are part of standard homeowner's or property owner's insurance policies.

Insuring Against Property Damage

For homeowners, the most cost-effective way to insure a new PV system against damage is under an existing homeowner's insurance policy. (This is far cheaper than trying to secure a separate policy for a PV system.) Businesses can cover a PV system under their property insurance, too.

When installing a PV system on a home or outbuilding — or even on a pole mount — contact your insurance company to determine if your current coverage is sufficient. If not, you'll need to boost the coverage enough to cover full replacement, including materials and labor. If you are installing a PV array on a pole, be sure to let them know that the PV system should be insured as an "appurtenant structure" under your current homeowner's policy. This is a term used by the insurance industry to refer to any uninhabited structure on your property not physically attached to your home. Examples include unattached garages, barns, sheds, satellite dishes, and wind turbine towers.

Insurance premiums on a homeowner's insurance policy fall into two different categories, each with differing rates. Your home is assessed at a higher value than an unattached garage or a storage shed. This is because people's homes are more lavish than most garages and sheds, and contain personal possessions, furniture, and clothing not typically found in other structures.

Appurtenant structures, are therefore charged at a lower rate. Insurance companies usually base premiums on appurtenant structures on the total cost of materials plus the labor to rebuild the structure.

PV systems should have insurance coverage that includes damage to the system itself from "acts of nature," a.k.a. "acts of God," plus possible options for fire, theft, vandalism, or flooding.

While most PV systems are designed to withstand 100-plus mile-per-hour winds, tornadoes or hurricanes can obviously destroy them, just as they would any other structure in their path.

Another "act of nature" of concern is lightning strikes. A properly installed PV system is properly grounded. Pole-mounted arrays are grounded at the base of the pole and the wire is grounded before it enters the home. Inside, the PV system ground is connected (bonded) to the electrical ground of the home's wiring. In addition, grid-connected systems are grounded on the utility side of the inverter.

Further protection is provided by installing lightning arrestors in appropriate locations. While lightning arrestors will not guarantee that your system is safe from lightning, it may reduce the damage from a rare direct or more likely nearby lightning strike. Plus, in the eyes of the insurance company, you have taken prudent measures to protect your system. If your system is fried by lightning, you'll be able to collect from the insurance company.

Fire is of minimal concern to a PV system, unless it is mounted on a home. Theft of a PV system, or a part of the system, is unlikely, unless the modules are mounted on the ground. Vandalism may also be a concern for ground-mounted PV systems. While incidents of vandalism are infrequent, they have occurred.

Flood insurance is a nationally administered program to protect primary dwellings. Costs can be exceptionally high for a home located along a coastline or in a floodplain near a stream or river with a penchant for flooding. If you live in a floodplain, obtain an insurance estimate before beginning construction.

Insuring a PV system is relatively inexpensive. While home insurance coverage should cover appurtenant structures, additional insurance may need to be purchased for an additional premium. Since most rural homeowner insurance claims are for fire damage, the deciding factor in pricing coverage in rural areas is determined by the distance of the home or business from the nearest fire department.

Liability Coverage

Liability insurance is also part of homeowners' or business owners' insurance policies. It protects against possible damage to others caused by a PV system. Liability insurance is required by the utility for homeowners and businesses that wish to install a utility-connected PV system. Liability insurance covers possible claims of damage to a neighbor's electronic devices from a grid-connected PV system. (For example, if your system sends power onto the line that somehow, magically, damages electronic equipment in a neighbor's home.) It also covers personal injury or death of employees due to electrical shock from a system when working on a utility line during a power outage. Even though the likelihood of this is nil — because of the automatic disconnect feature built into inverters — utilities still insist on this coverage.

Liability insurance in the amount of $100,000 is considered adequate for small PV systems by most utilities. In most places, liability coverage for homes runs from $100,000 to $300,000.

In addition to liability insurance, utilities may require you to indemnify them from potential damage caused by your PV system.

Buying a PV System

If a PV system seems like a good financial, social, or recreational pursuit for you, I strongly recommend hiring an experienced, competent, and trustworthy professional installer. Local suppliers/installers with experience and knowledge can be a great ally. They can supply all of the equipment, be certain that it is compatible, obtain or assist with permits and interconnection agreements, and, then, of course, install the system. They will test it to be sure it is operating satisfactorily. They should be there to answer questions and to address problems you have with the equipment.

Most PV modules come with a 20- to 25-year production warranty, and grid-tied batteryless inverters are often guaranteed for five years. Many manufacturers offer extended warranties, most notably when a ten-year warranty is required to qualify for special installation incentives. Battery-based inverters are often guaranteed for two to five years, but again, an extended warranty may be available. Reliable equipment and local installers who know the business are both worth their weight in gold. Look for people who've been in the business for a while and who have installed a lot of systems. The longer the better; the more systems the better.

Be sure your installer is bonded, too. A bond is a sum of money set aside by contractors and held by a third party, known as a *surety company*. The bond provides financial recourse to homeowners and business owners if a contractor fails to meet his or her contractual obligations. In such instances, customers can file a claim for compensation from the bonding company. To stay bonded, a contractor must reimburse the company to cover any payments made to customers. Residential contractors usually carry a minimum of $10,000 bond and some AHJs require that contractors be bonded.

When contacting companies, ask if they are bonded and the amount of bond. Check it out. Also ask for references, and be sure to call them. Visit installations, if possible, and contact the local office of the Better Business Bureau. Get everything in writing. Sign a contract. Be sure the installer has worker's compensation insurance to protect his or her employees when working on your site.

Don't pay for the entire installation up front. Be sure you are on the site when the work is being done. The accompanying sidebar, "Questions to Ask Potential Installers," includes a list of questions you should ask when shopping for an experienced and competent installer.

You can also purchase equipment from a local supplier and install it yourself provided the AHJ allows property owners to perform electrical work. If you live in such an area, you can file for permits and install all equipment — PV arrays and electronics. However, you cannot legally hire any unlicensed electricians to act as the electrical contractor or employ any unlicensed individuals to work on the system. You have to do the work yourself, except for the final grid connection, which must be performed by a licensed electrical contractor. The AHJ may also require the owner/installer to live in or occupy the building and may restrict him or her from selling or leasing the property for one year.

As a rule, I don't recommend this route unless you are handy and you have attended a couple of PV installation workshops like those taught through my organization, The Evergreen Institute, (www.evergreeninstitute.org) or similar organizations like Solar Energy International, the Solar Living Institute, and the Midwest Renewable Energy Association.

Installing a PV system is risky. Working on a roof and wiring are fraught with difficulties. Connecting to the electrical grid is a job for professionals. Even professional PV installers must hire licensed electricians to do the work. A PV system is a huge investment and you don't want to mess it up.

You can also purchase equipment online from manufacturers or from online dealers. Some vendors sell complete packages with PV modules, inverters, racks, and everything else. However, be aware that getting help installing packaged systems from such suppliers can be difficult.

Questions to Ask Potential Installers

1. How long have you been in the business? (The longer the better.)

2. How many systems have you installed? How many systems like mine have you installed? (The more systems the better.)

3. How will you size my system?

4. Do you provide recommendations to make my home more energy efficient first? (As stressed in the text, energy efficiency measures reduce system size and can save you a fortune.)

5. Do you carry liability and worker's compensation insurance? Can I have the policy numbers and names of the insurance agents? (Liability insurance protects against damage to your property. Worker's compensation insurance protects you from injury claims by the installer's workers.)

6. Are you bonded? For how much, and with whom? (Bonding, as explained in the text, provides homeowners with financial recourse if an installer does not meet his or her contractual obligations.)

7. What additional training have you undergone? When? Are you NABCEP certified? (Manufacturers often offer training on new equipment to keep installers up-to-date. NABCEP is a national certifying board that requires installers to pass a rigorous test and have a certain amount of experience.)

8. Will employees be working on the system? What training have they received? How many systems have they installed? Will you be working with the crew or overseeing their work? If you're overseeing the work, how often will you check up on them? If I have problems with any of your workers, will you respond immediately? (Be sure that the owner of the company will be actively involved in your system or that he or she sends an experienced crew to your site.) ☞

9. Are you a licensed electrician or will a licensed electrician be working on the crew? (State regulations on who can install a PV system vary. A licensed electrician may not be required, except to pull the permit, supervise the project, and make the final connection to your electrical panel.)

10. What brand modules and inverters will you use? Do you install UL-listed components? (To meet code, all components must be UL listed or listed by some other similar organization.)

11. Do you guarantee your work? For how long? What does your guarantee cover? How quickly will you respond if troubles emerge? (You want an installer who guarantees the installation for a reasonable time and who will fix any problems that arise immediately.)

12. Do you offer service contracts? (Service contracts may be helpful early on to be sure the system runs flawlessly.) How much will a service contract cost? What does it cover?

13. Can I have a list of your last five projects with contact information? (Be sure to call references and talk with homeowners to see how well the installer performed and how easy he or she was to work with.)

14. What is the payment schedule? Can I withhold the final 10% of the payment for a week or two to be sure the system is operating correctly? (Don't pay for a system all at once. A deposit, followed by one or two payments protects you from being ripped off. Never make a final payment until you are certain the system is working well.)

15. Will you pull and pay for the permits?

16. Will you work with the utility to secure an interconnection agreement?

17. What's a realistic schedule? When can you obtain the equipment? When *will* you start work? How long will the whole project take? ∎

Parting Thoughts

When you started reading this book, you no doubt were already interested in PV systems. Perhaps you just wanted to determine if a PV system would be suitable for your home. Perhaps you were sure you wanted to install a PV system but didn't know how to proceed. We hope that you now have a clear understanding of what is involved.

If you have come to realize that your dreams for a PV system were not realistic, be glad that you did not spend a lot of money on a system that would not have met your expectations. If, however, you now have an informed conviction that a PV system is for you, you may want to locate a dealer/installer.

You are now up to speed on photovoltaic systems. You know a great deal about solar energy, electricity, PV modules, PV systems, inverters, batteries, charge controllers, and permits. If you want to learn more, I urge you to sign up for workshops and check out a copy of my more comprehensive *Power from the Sun*. Whatever you do, be safe and enjoy the journey and let the Sun shine in!

Index

About the Author

D an Chiras is an internationally acclaimed author who has published over 24 books, including *The Homeowner's Guide to Renewable Energy* and *Green Home Improvement*. He is a certified wind site assessor and has installed several residential wind systems. Dan is director of The Evergreen Institute's Center for Renewable Energy and Green Building (www.evergreeninstitute.org) in east-central Missouri where he teaches workshops on small wind energy systems, solar electricity, passive solar design and green building. Dan also has an active consulting business, Sustainable Systems Design (www.danchiras.com) and has consulted on numerous projects in North America and Central America in the past ten years. Dan lives in a passive solar home powered by wind and solar electricity in Evergreen, Colorado.

If you have enjoyed *Solar Electricity Basics* you might also enjoy other

Books to Build a New Society

Our books provide positive solutions for people who want to make a difference. We specialize in:

Sustainable Living • Green Building • Peak Oil
Renewable Energy • Environment & Economy
Natural Building & Appropriate Technology
Progressive Leadership • Resistance and Community
Educational and Parenting Resources

New Society Publishers

ENVIRONMENTAL BENEFITS STATEMENT

New Society Publishers has chosen to produce this book on Enviro 100, recycled paper made with **100% post consumer waste**, processed chlorine free, and old growth free.

For every 5,000 books printed, New Society saves the following resources:[1]

15	Trees
1,357	Pounds of Solid Waste
1,494	Gallons of Water
1,948	Kilowatt Hours of Electricity
2,468	Pounds of Greenhouse Gases
11	Pounds of HAPs, VOCs, and AOX Combined
4	Cubic Yards of Landfill Space

[1]Environmental benefits are calculated based on research done by the Environmental Defense Fund and other members of the Paper Task Force who study the environmental impacts of the paper industry.

For a full list of NSP's titles, please call 1-800-567-6772
or check out our website at: **www.newsociety.com**

NEW SOCIETY PUBLISHERS